Ocean Container Transportation

AN OPERATIONAL PERSPECTIVE

Mark L. Chadwin
James A. Pope
Wayne K. Talley

Taylor & Francis
New York • Bristol, PA. • Washington, D.C. • London

USA	Publishing Office:	Taylor & Francis New York Inc. 79 Madison Ave., New York, NY 10016-7892
	Sales Office:	Taylor & Francis Inc. 1900 Frost Road, Bristol, PA 19007
UK		Taylor & Francis Ltd. 4 John St., London WC1N 2ET

Ocean Container Transportation: An Operational Perspective

Copyright © 1990 Taylor & Francis New York Inc.

All rights reserved. No part of this publication may be reproduced, stored in a retrieval system, or transmitted, in any form or by any means, electronic, electrostatic, magnetic tape, mechanical, photocopying, recording or otherwise, without the prior permission of the copyright owner.

First published 1990
Printed in the United States of America

Library of Congress Cataloging in Publication Data

Chadwin, Mark Lincoln
    Ocean container transportation: an operational perspective/Mark
L. Chadwin, James A. Pope, and Wayne K. Talley.
    p. cm.
    Bibliography: p.
    Includes index.
    ISBN 0-8448-1628-0
    1. Containerization. 2. Shipping. I. Pope, James A. (James
  Arthur), 1950–   . II. Talley, Wayne Kenneth. III. Title.
TA1215.C47   1989
387.5'442—dc20   89-34094
                                                                                   CIP

Contents

Preface v
Acknowledgments vii
Chapter 1. The Advent of Containerization 1
Chapter 2. The Marine Container Terminal and Its Operation 19
Chapter 3. Costing Terminal Operations and Measuring Capacity 35
Chapter 4. Evaluating Terminal Operations 57
Chapter 5. Ship Technology and Costing Containership Service 79
Chapter 6. Networks, Intermodalism, and Containership Service 93
Chapter 7. Ocean Container Transport in the Future 111
Index 137
About the Authors 141

Preface

Beginning in 1983, the College of Business and Public Administration at Old Dominion University in Norfolk, Virginia, undertook a program of applied and scholarly research in maritime transportation and trade. Over the ensuing five years, the Research Fellows in the program conducted a wide variety of projects involving marine terminal operation, intermodal system development, and ocean carrier management.

Much of this work focused on the problem of optimizing efficiency in the movement of ocean containers. It included:

Time and motion studies of on-terminal container handling.
Analyses of the operational and organizational implications of emerging technology such as high-speed cranes, unit trains, and computerization.
Economic impact studies of port activities that estimated benefits in terms of jobs, incomes, and tax revenues.
Design of a model to allocate the costs associated with container terminal operations.
Analysis of container ship operating costs.
Development of procedures for assessing costs and benefits of alternative routing strategies, such as load centering, multiporting, and feeder services.
Conducting an international symposium for the Federal Maritime Commission on the impacts of the Shipping Act of 1984 on carriers, shippers, and ports.
Analysis of the impacts of road and rail infrastructure on marine terminals, and visa versa.
A study of innovations in labor relations and personnel management at marine terminals.
Research on the effects of landbridging and on the customer base for an inland container collection facility.

Each of the authors was a major participant in that program. As the College's director of research and a professor of international business and organization design, Mark Chadwin developed and directed the program. Wayne Talley, now chairman of the Department of Economics and an Eminent Scholar in transportation economics, served as a Research Fellow of program. So did James Pope, who currently chairs the college's Department of Management Information Systems/Decision Sciences. Although this volume draws heavily on work originally undertaken during the program, developments in the field continue to move rapidly. Therefore, a good deal of new material has been included.

Chapter 1 examines the origins of containerization and the forces that have molded its development through the 1980s. It also provides an overview of the ways in which countainership operators, ports and marine terminals, and other actors in the transportation system have sought to respond to those forces. Ensuing chapters focus on one particularly important set of responses, namely those that improve efficiency and reduce costs. What is being done—and what can be done—to minimize cost and increase productivity?

The book is organized into three parts. The first focuses on marine terminals. Chapter 2 explains how a marine container terminal functions and describes the equipment, individuals, and organizations that may be involved in its operation. It emphasizes the management of information about the containers as well as the manner in which they are handled and stored.

Chapter 3 presents techniques for identifying and allocating the cost of container terminal development and operation to the different types of actors involved in the terminal. This chapter also examines the strengths and weaknesses of various methods for estimating the capacity of container terminals.

Chapter 4 describes the operational problems that typically confront the managers of a container terminal and presents a variety of techniques with which to evaluate a facility's performance. These techniques are intended to help identify the sources of operational problems and analyze alternative solutions to them.

The second part of the book concentrates on the containerships themselves. Thus Chapter 5 examines the vessel operator's problem of costing containership service. It presents a cost analysis of ocean containership investment and operation, and it considers such issues as optimal containership size and route distances.

Chapter 6 addresses the problem of choosing between alternative routing patterns or interport networks. In the process, it examines such operating strategies as the use of feeder services, load centering, all-water movement, and intermodal movement. The chapter also examines the impact of transportation deregulation in the United States on carriers' choice of networks.

The third section consists of one chapter that considers ocean container transportation and intermodalism during the 1990s and beyond. Thus it examines the likely pace of containerization, the characteristics of future containerships and container terminals, and the implications of emerging technologies, commercial patterns, and government policies. Finally, it speculates on the nature of the global transportation providers of the future and the factors that will influence their success or failure.

Acknowledgments

The authors wish to express our gratitude to the many individuals and organizations that helped make this book possible. They include the operators of container terminals and the staff of port authorities in Antwerp, Charleston, Felixstowe, Hamburg, Hampton Roads, Hong Kong, Koper, Liverpool, New York-New Jersey, Ningbo, Rotterdam, Savannah, and Zeebrugge, who have provided us with information, ideas, and hospitality.

We would like to thank those organizations that have assisted in our research or allowed us to work with them on various projects over the past decade, namely ECT-Rotterdam; Evergreen Marine Corporation; the Federal Maritime Commission; the Hampton Roads Terminal Operators' Association; Lamberts Point Docks, Inc.; Nippon Yusen Kaisha (NYK) Line; the Office of Port and Intermodal Development, U.S. Maritime Administration; the Virginia Center for Innovative Technology; Virginia International Terminals, Inc.; and the Virginia Port Authority. We are particularly grateful in this regard to our home institution, Old Dominion University, for the support provided for our Maritime Trade and Transport program as well as for the preparation of this manuscript.

Also, we wish to acknowledge our colleagues and friends who read and commented on parts of this manuscript—Robert S. Agman of the Labor-Management Maritime Committee, J. Robert Bray of the Virginia Port Authority, Edward Brown of the International Longshoremen's Association, Adrienne Chadwin, Philip A. Dieffenbach of Norfolk Southern Corporation, Joseph Dorto of Virginia International Terminals, Inc., Richard O. Goss of the University of Wales at Cardiff, Dominic Obrigkeit of Evergreen Marine Corporation, Eric Pollock of ABP Consultancy, Ltd., and Hiroshi Takahashi of NYK Lines. Although we are indebted to them for their time and thoughts, the responsibility for any shortcomings is ours alone.

Mark L. Chadwin
James A. Pope
Wayne A. Talley

Chapter 1

THE ADVENT OF CONTAINERIZATION

THE BEGINNINGS

Man has been experimenting with containers since the dawn of commercial history. The merchants who first sought to improve cargo handling and protection by placing two small parcels in the same crate or using sealed *amphorae* took the earliest steps toward containerization as we know it today.[1] Over the centuries other attempts were made to simplify cargo movement and consolidate shipments into larger, standardized parcels. However, these efforts usually were defeated by limitations in the technology of cargo handling and movement.

Some advances occurred in the handling of bulk commodities when casks and barrels were replaced by specially designed ships into which oil, coal, or grain could be poured. However, there was relatively little progress with the rest of international commerce, so-called general cargo. Through the 1950s, it continued to be handled "break-bulk" style. This meant the movement of freight, generally one parcel at a time, onto a truck or rail car that carried it from the factory or warehouse to the docks. There each parcel was unloaded and hoisted by cargo net and crane off the dock and onto the ship. Once the package was in the ship's hold, it had to be positioned precisely and braced to protect it from damage during the ocean crossing. This process was performed in reverse at the other end of the voyage. Thus the movement of ocean freight was slow, labor-intensive, and expensive.

McLean's Revolution

All of this began to change in 1955. Malcom McLean, the owner of a North Carolina trucking firm, had long believed that individual pieces of cargo needed to be handled only twice—if they could be packed at the factory into a truck trailer and if the *entire trailer* was then moved to the seaport, across the ocean, and to the door of the recipient. Only then would the trailer be unloaded. McLean was aware that some railroads were already moving trailers full of cargo across country on flat cars. He purchased a small ship line, renamed it Sealand, and began experimenting with the movement of trailer loads of cargo from New York to the Gulf Coast and Puerto Rico in 1956. For McLean it was only a logical extension to load several hundred trailers onto a refitted World War II tanker in the spring of 1966 and send them across the Atlantic.

The containerization of international commerce had truly begun. In the years that followed, standardized trailer bodies were constructed, generally 20 or 40 feet long, eight feet wide, and eight or eight and a half feet high. One 40-foot container could accommodate 20 tons of smaller parcels. Each container had locking mechanisms at each corner that could be secured to a truck chassis, a rail car, a crane, or to other containers inside a ship's hold or on its deck.[2]

Intermodalism

The use of standardized containers meant that "intermodalism," the movement of goods from point to point by more than one mode of carrier, became commercially feasible. In the strictest sense intermodalism had existed from the time that cargo had been drayed by horse cart to the dock and thence taken by sailing ship to its destination. Now, however, the movement of high volumes of freight at unprecedented speed across both land and water became practicable.

To accommodate the changes, new investments were needed on both land and water. Gradually, the refitted tankers and breakbulkers that first carried 200–300 containers were replaced with vessels designed specifically to accommodate a thousand or more of these standardized "boxes." On land, new terminal equipment was necessary—shipside cranes that could load and unload the containers faster than the ships' cranes and specialized equipment for moving the containers around the yard.

Furthermore, the terminals themselves had to change. Traditionally, they had consisted of rows of finger piers that jutted out into the harbor. Many of the piers were covered with sheds to protect the breakbulk cargo. Now, instead, long stretches of bulkheaded waterfront were needed along which the big cranes could move as they worked the ship and, next to the cranes, areas of open space in which the containers and their chassis could be quickly moved and stored. Modifications of traditional truck chassis, railcars, and inland barges also were necessary to move the containers farther inland.

Even in the advanced industrialized countries of Western Europe and North America, these transitions occurred only gradually over the next decade and a half. Furthermore, full exploitation of the potential of containerization and intermodalism had to await other developments—in the global economy, information technology, and governmental policy.

Nevertheless, "the industrial revolution in international shipping" had begun.[3] As in all revolutions, important new actors emerged and some traditional ones slipped into decline or disappeared altogether. Thus major new ship lines, like Sealand, emerged, while some that were slower to adapt were merged or dissolved. Similarly, the cargo-handling activities of the ports themselves gradually moved out of the crowded downtown settings of finger piers and warehouses into more open space on the outskirts of the big cities.

The Potential Benefits

All this investment was expensive, but the advantages to the ocean carrier and the shipper in labor, time, and other costs were enormous. Previously, the carrier had to employ a gang of perhaps 20 longshoremen to load each hatch on a breakbulk vessel. Each such gang could load perhaps 20 tons per hour. Now it was possible to load one container whose contents weighed that much in two or three minutes. One crane and perhaps half as many men could load and stow 400 or 500 tons an hour. As a result, the time a vessel spent in port could be greatly reduced. Where breakbulk freighters often took a week to unload and reload, a container ship might call for only four to six hours. Less time in port meant not only lower costs for dockage and wharfage, but also faster turnaround for each vessel. Faster circuits meant that fewer vessels were needed to carry the same amount of cargo between the same set of ports in a given time period.

For the shipper, less handling meant less frequent damage to the cargo. Since containers could be sealed at the warehouse or plant and not opened until they arrived at the customer, there was less loss from pilferage, long an accepted cost of shipping. Lower losses of both types meant lower costs in insurance premiums. Furthermore, packaging did not have to be so sturdy, and the expense of building customized crates could be avoided. Lighter packages cost less to transport. Finally, delivery of goods was not only faster, but it also was more reliable. It was possible for a container load of consumer products to reach port one day and be at the retail outlet the next. As certainty about the date of delivery improved, inventories could be cut back substantially, generating further savings. Furthermore, it was now sometimes possible to obtain shipping rates by the "box" rather than by the commodity, a considerable advantage to the shipper of higher value goods. In theory at least, all these cost efficiencies contributed to the expansion of international trade and to lower prices for the consumer. By 1988 the ocean transportation cost from Japan to the United States for a VCR retailing for $200 was less than $2.[4]

IN THE EIGHTIES—THE CONTEXT

By the mid-1980s the containerization-intermodalism revolution was in full cry, augmented by changes in the economic, technological, political, and labor environment.

Economic Changes

Postwar economic recovery and a sustained period of economic expansion brought high rates of growth in the value and volume of trade, especially among the industrialized countries. During the two decades from 1960 to 1980, the world's gross product grew at an average annual rate of about 5 percent.[5] International commerce, abetted by declining transportation costs and a series of tariff reductions negotiated under the General

Agreement on Trade and Tariffs, more than kept pace. In current dollar terms, world trade grew from $128 billion in 1960 to nearly $2 trillion in 1980, an annual increase of 11.5 percent. During the same period, export of manufactured products, the primary market for containerized transport, expanded more than 15 times in value, rising from $65 billion in 1960 to $1,095 billion in 1980.[6] Containerization both contributed to and matured in a period of global trade growth unique in human history.

High interest rates and new patterns of production and distribution in the late 1970s and 1980s reinforced the impetus toward containerization and intermodalism. High interest rates made large inventories prohibitively costly. Just-in-time (JIT) inventory management techniques first applied in Japan minimized these costs, and these methods were adopted by producers, distributors, and retailers in the United States and other industrialized countries. At the same time, multinational enterprises, seeking to minimize production costs, created global networks of interdependent manufacturing and assembly facilities, each of which depended on the timely arrival of materials from the others. Furthermore, retailers sought to adopt strategies of "quick response" that permitted them to react to shifts in fashion and taste by having newly designed merchandise in the stores in a few weeks. Such practices put new pressures for speed and reliability on each link in the transportation system—the ocean carrier, the marine terminal, the railroad, and the truck line.

Rapid and large swings in key foreign exchange rates during the 1980s made the environment even more turbulent. The U.S. dollar, the central currency of world trade since the end of World War II, nearly doubled in strength against the pound, the French franc, and the Deutschmark from 1979 to 1985, before falling nearly as far in value in the next two years. The effects were dramatic shifts in industrial production and major trade flows. Each shift tended to create new cargo imbalances, posing operational and financial problems for ocean carriers.

As the dollar strengthened, U.S. export flows slowed, and imports from Europe, Japan, and the newly industrialized countries of Asia and the Pacific accelerated. Vessels sailing eastward across the Pacific or westward across the Atlantic went full, but on the return leg they carried empty containers or low value (and low revenue) cargo like scrap and waste paper. Rates, routing patterns, equipment positioning, port facility investment, and marketing strategies had to be rethought and revised. The profitability of some U.S. container lines, already losing ground as a group to lower cost Asian operators, was damaged further by the strong dollar. This, together with excess capacity and declining rates, contributed to a series of consolidations and bankruptcies.

Then, as the industry began to adjust to these conditions, the situation changed again. The dollar fell dramatically, and cargo flows began to reverse. Imbalances in cargo and capacity began to appear in the opposite direction. Outbound voyages from the United States were more fully laden and profitable, while incoming vessels were often partly full. As the yen nearly doubled in value against the dollar, Japanese liner companies in particular suffered large financial losses. Furthermore, the rising value of both the won and yen foretold of price increases for new containerships and containers as well as shifts in sources of supply.[7]

Government Policy

Governmental action, particularly in the regulatory arena, exerted a major influence. Much attention was focused in the late 1970s on the implications of supranational regulation, especially the U.N. Conference on Trade and Development's Liner Code and its rigid cargo-sharing formula.[8] However, it was transportation deregulation in one country, the United States, that had a far more immediate effect. Beginning in 1978, legislation was passed that relaxed regulatory controls over airlines, trucking and railroads, and—in 1984—ocean shipping.

This body of legislation diminished the role of government in decisions about price and service, while giving legal sanction to commercial arrangements that crossed transportation modes. Heretofore, carriers charged each shipper a uniform, common carrier rate that was recorded in publicly available tariffs. The Shipping Act of 1984, however, permitted shippers and carriers collectively through conferences to negotiate lower rates and to engage in service and time-volume contracts without fear of antitrust or regulatory action. The act also granted permission for negotiations between liner conferences and inland carriers about rates and service and "point-to-point" rates for intermodal services from origin to destination.

The new legal climate offered opportunities to exploit more fully the potential of containerization and intermodalism, while reinforcing the competitive pressures to do so. Behavior was more market-driven. Increasingly, the customer called the tune rather than the regulator. Carriers bid for the business of the shippers—especially the large ones—and ports and marine terminals bid for the business of the carriers. "Customer service" became fashionable throughout the transportation industry.[9]

If government deregulation facilitated the development of containerization and intermodalism, the performance of some traditional government functions complicated that development. For example, inspections of containerized cargo—sometimes intensified due to illicit drugs, counterfeit goods, or terrorist threats—caused delays that became particularly troublesome in the tightly scheduled world of JIT and intermodalism. Since inspection now sometimes meant "devanning" a whole container at the shipper's expense rather than opening a single crate, it was also much more expensive.

Conversely, governments, like the industry itself, were moving to computerize—to automate data collection, customs entries, clearances, and other procedures. This was intended to increase the efficiency and speed of cargo movement.

Information Technology

With intermodalism and JIT, information about the cargo—its location, clearance status, and arrival time—often seems as important as the cargo itself. Yet, ocean shipping was slower than other industries to apply the managerial and operational capabilities of the computer and advanced electronic communications technologies.[10] The computer *was* sometimes applied to vessel design, officer training, or even stowage planning. Generally, however, utilization was impeded by traditional preferences for hard copy documentation (often hand carried by messengers), use of documents as negotiable instruments, fears about information confidentiality, and concerns about the cost and reliability of the unfamiliar new technologies.

This changed drastically in the late 1980s. All types of carriers, as well as marine terminals, port authorities, and other intermediaries, moved to computerize their own operations and to exchange data electronically with each other and with shippers and consignees. The pressures for interorganizational and interindustry compatibility led to a wide variety of standardization efforts—generic electronic formats for all sorts of documents, unique identification numbers for each shipment, a harmonized commodity code, and adoption of universally accepted terminology and definitions. This development occurred amid continuing concern about the viability of some traditional intermediaries (smaller freight forwarders, customs brokers, and messenger services) as well as about the access of unwelcome "outsiders" to commercially sensitive information. The transition to the new technologies also sometimes caused periods of chaos and reduced efficiency. However, the end result was an increase in the speed and precision with which both the cargo and information about it moved.

Labor

If information technology generally enhanced intermodalism and containerization, industrial relations often were impediments. As noted earlier, containerization radically altered cargo handling tasks. Far fewer workers were needed, and the skills requirements were different. Instead of men with manual skills and physical strength, the priority was on two types of personnel—operators and maintenance workers for the sophisticated and expensive heavy equipment that handled the containers, and various white collar workers to collect, transmit, and use information about the containers. Longshore unions had struggled for nearly a century to win favorable wages, benefits, and economic leverage. They were understandably reluctant to accept the changes.[11]

Different sorts of accommodations were reached in different places. Sometimes work rules, gang sizes, and compensation patterns remained pretty much as they had for handling breakbulk, under various "work preservation" schemes. On the Atlantic and Gulf coasts of the United States, labor-management negotiations led to a "50-mile rule" (reserving all "stuffing and stripping" of containers in or near port cities for unionized longshoremen), guaranteed annual incomes regardless of hours actually worked, and agreements that required ocean carriers to use union labor wherever their vessels called. The resulting costs were passed on to the shipper, and ultimately the customer.

In other cases compromises were reached that involved productivity-related pay schemes, early retirement programs, gradual attrition, or retraining in the new technology. In still other situations, terminal operators who used nonunion labor (or less expensive workers from other unions) offered carriers lower cost and more responsive services, undercutting nearby unionized facilities.

Despite the enormous growth in general cargo tonnage, jobs declined dramatically along the docks. In 1970 30,000 longshoremen in the port of New York-New Jersey worked 33 million manhours to move about 25 million "assessment tons." In 1986 there were only 7,400 dockworkers on the rolls, and they worked a little more than 11 million manhours to move about the same amount of cargo.[12] In the United Kingdom, dock jobs fell from 80,000 in 1967 to 11,400 in 1986.[13] By the late 1980s the pressures of interport competition and the resulting need for efficiency and cost-consciousness were causing organized labor to accept some of the operational and workforce implications of containerization, even where they had been slow to do so in the past.

Similarly, work and life aboard ship changed dramatically with containerization. Port calls of several days, which were part of the allure of the life of the merchant seaman, were replaced by stops that often lasted only four to six hours. Crews could spend months aboard without setting foot on land. To offset this, some carriers improved onboard living quarters and added small swimming pools and gymnasia to some of their vessels.

As vessel designs changed, fewer merchant seaman were needed to operate ever-larger containerships. Automated engine rooms, advances in piloting and information systems, navigation satellites, and centralized control from global corporate headquarters via high-speed telecommunications also altered the responsibilities of masters and other officers. By 1986 experts were forecasting containership crews of as few as a half-dozen, and in 1988 one Swedish company began operating a combination container-Ro-Ro (roll-on roll off) ship in transAtlantic service with a crew of only 12.[14]

IN THE EIGHTIES—THE OCEAN CARRIERS

Restructuring of the Industry

The restructuring of the general cargo sector, which began with the advent of containerization, accelerated in the 1980s. Following the pattern in the bulk trades, investors from the newly industrialized countries (NICs) of Asia and the Pacific—Taiwan, Hong Kong, Singapore, and Korea—emerged as major containership operators. By 1988 Asian operators controlled about 30 percent of the world container fleet, and the proportion operated by Western Europeans and North Americans had declined to about 50 percent.[15] Important participants included Hanjin, Yang Ming, OOCL, Neptune, and especially Evergreen, which by 1987 had become the largest ocean carrier of containers. As relatively late entrants, they often had the advantage of newer, more cost-efficient and larger vessels. Other potentially significant competitors included ship lines representing nonmarket economies, such as Polish Ocean Lines and the China Ocean Shipping Company, some of which had the advantages of captive cargoes and less pressure for profitability.

Firms that had previously been important continued to disappear, merge, and consolidate, or invested heavily in distinctive strategies to survive. Examples abounded. United States Lines went out of business in 1987. Peninsula and Orient (P & O) acquired Trans Freight Lines and Overseas Container Lines and negotiated a space-sharing agreement with Nedlloyd. Then Sealand, traditionally an independent actor, entered into slot-sharing and joint services arrangements with TFL and Nedlloyd. Japan's five principal liner companies, which had earlier been consolidated from a dozen separate firms, suffered several years of increasingly losses and were pressured by their government into further consolidations.

American President Lines took the lead in defining a strategy based on intermodalism. They invested in containerships too big for the Panama Canal that would operate in the Pacific only and in a network of double-stack trains to carry cargo from the U.S. West Coast to major markets across the country. By the end of 1988, Japanese, Taiwanese, and Korean carriers also were investing heavily in intermodal strategies in the North American market.

The ship lines that had survived were beset with concerns about excess capacity, a traditional problem in maritime industries. During this century, demand for carrying capacity globally has exceeded supply only rarely—usually in wartime. Tonnage built during a boom or war might survive two decades, causing depressed rates and profitless operation. When demand strengthened and rates firmed a bit, operators would often replace their aging fleets with more modern, larger vessels. Frequently, however, the older vessels would not be scrapped but would be bought and operated by someone else.

In the late 1970s and early 1980s, banks made excessive loans for new vessel construction. If a line went bankrupt, control of its ships would often revert to the bank, which would sell or charter them to someone else. National governments and shipyards hard pressed for business also gave excessively generous terms for new vessel construction, often subsidizing the transaction with public funds. In some countries, nationalism and employment considerations led to the continued operation of national shipping lines, even though they were no longer economically viable. By the late 1980s estimates of worldwide containership overcapacity ranged from 20 to 30 percent.[16]

Excess capacity meant increasing price competition. In the crucial U.S. routes, provisions of the Shipping Act of 1984 that permitted service contracts and required liner conferences to allow "independent action" on pricing by each of their members exerted further downward pressure on rates. These provisions also had the effect of weakening the conferences' control over both rates and service.[17]

The Search for Responses

Carriers responded to excess capacity by "rationalizing" services, often in collaboration with their competitors. Thus joint services, slot sharing, feeder services, shared on-land facilities, and container or chassis pooling increased. Another response, more traditional among bulk carriers, involved using open registries (flags of convenience such as Panama or Liberia) or newly created "off-shore" or "dual" registries. These mechanisms lowered manning costs by permitting operation with smaller crews or with foreign seamen and avoided other expenses associated with home country registry.

Increased pressure for competitiveness influenced the lines' marketing efforts. Through their sales representatives and advertisements, they sought to differentiate their products by selling speed (via mini-landbridge),[18] reliability (fixed day service), expertise in par-

Advent of Containerization

ticular markets or cargoes, customer service (point-to-point rates and special handling), as well as choices of price and level of service (mini-landbridge vs slower but less expensive all-water shipment). In addition, they sought entirely new cargoes, often of lower value, to containerize.

Some of the decisions by individual companies caused radical shifts in the context in which both they and their competitors operated. Tonnage in Far East-North American trades more than doubled in the five years from 1983 to 1988, as firms overbuilt or switched tonnage to the fastest growing market. Similarly, tonnage in the North Atlantic trades increased by about 25 percent in a few months when Maersk decided in 1988 to enter this market and Sealand bought and deployed the huge former U.S. Lines "Econships" on the same routes.

Faced with declining profitability, a number of carriers diversified out of shipping and even transportation. By the late 1980s, the revenues and profits of a number of shipping companies were coming more from their onshore investments than from ocean carrier operations. Others became involved (voluntarily or otherwise) in mergers or buyouts by larger transportation firms that saw potential synergies (like CSX-Sealand) and had the financial resources to sustain the struggle.

Among those that stayed in the industry, different carriers responded to the increasingly dynamic and complex environment by adopting different strategies. Evergreen pursued a strategy based on round-the-world service, using very large vessels in continuous westbound and eastbound circuits. Although this concept had been forecast to be the "wave of the future," by the late 1980s only two other carriers, Nedlloyd and Senator Lines, sailed individual vessels around the world, and their strategies were very different from Evergreen's. Nedlloyd used small vessels capable of carrying noncontainerized cargo to call on Southern Hemisphere and U.S. West Coast ports only. Senator chartered the vessels they operated, generally called at small nonunion ports, offered port-to-port services only, and operated entirely outside of conferences. Other carriers offered round the world service but provided it by landbridging across the United States or by transshipping via their own vessels or someone else's.

Some carriers stuck largely to the big volume (and highly competitive) "mainstream" routes between North America, Europe, and Asia. Others focused on less developed and perhaps less competitive routes involving north-south flows. Many blended both or altered their routes frequently as they searched for a profitable niche.

Vessel operators made other diverging choices. Some "load centered" by calling at only one or two ports on a range. Others followed "multiporting" strategies, stopping at half a dozen places along the same coast. Some, like Sealand, sought to provide full service from door-to-door utilizing their own facilities and equipment. Others also offered comprehensive service, but utilized unrelated firms to provide rail, truck, or other intermediary services. Still others offered water transport only. This last group included some specialized or proprietary carriers of commodities such as fruit, lumber, or steel, which might carry containers on their backhauls to avoid sailing empty.

Landside patterns varied, too. Some carriers owned or leased their own terminals. Others utilized publicly owned, common-user terminals. Some ran their own exclusive unit trains, barges, or feeder services. Others relied on common carriers. Some were

moving their containers directly between the ship and over-the-
_____ speed and flexibility. Others put their boxes in "stacks" at the
_____ pace and minimizing their investment in chassis.

Containership Fleet

In the early 1980s some experts predicted that economies of scale would drive containership design quickly in the direction of very large vessels that were "nonself-sustaining" or "gearless." Such vessels, having no cargo cranes of their own and no roll-on roll-off ramps, would have to call at marine terminals equipped with big dockside container cranes. These "jumbo" ships would have capacities of 3,000 to 5,000 TEUs (20-foot equivalent units). The expectation was that they would call at only a few ports where large volumes of containers would be concentrated by ground transportation, barges, and small feeder vessels. These so-called load centers were analogous to the "hubs" in air carriers' "hub and spoke" routing patterns.

In reality, strategy drove ship design, and vice versa. Carriers whose strategies were based on load centering or calling only at ports with high volumes of cargo usually did opt for larger, nonself-sustaining containerships. However, those that called at smaller ports or were particularly concerned about flexibility in the uncertain future continued to invest in smaller vessels that were able to load and discharge their own containers. These included self-sustaining or "geared" containerships, RO-RO vessels, and various type of "semicontainer" or combination ships. Similarly, a few carriers such as APL purchased vessels that were more than Panamax size—too large to transit the Panama Canal—and, thus, constrained to operate in only one ocean.

Design decisions involving hundreds of millions of dollars in construction costs often hinged on forecasts of the unknowable—not only whether but also *when* global trade patterns would change or fuel oil prices would rise and fall. Even the giants of the industry could guess wrong. Thus McLean, who seemed so prescient in 1955, mistimed events 30 years later when he invested half a billion dollars in the fuel-efficient but slow Econships. When fuel oil prices collapsed, problems with the Econships helped drive United States Line into bankruptcy.

The world container fleet continued to be quite varied in type and size in the late 1950s. Although a trend to larger, gearless vessels was discernible, the transition seemed far more gradual and less widespread than had been forecast earlier.[19]

IN THE EIGHTIES—PORTS AND TERMINALS

Pressure from the Carriers

The ports and terminals felt the impact of the commercial, technological, and regulatory trends of the 1980s largely through the expressed needs of their primary customers, the ocean carriers. As indicated earlier, the overall effect of these trends was to reinforce the turbulence and competitive pressures felt by the carriers and drive them to search for new cost, price, and service strategies that offered the prospect of survival and profitability. The carriers, in turn, transmitted many of these pressures to the marine terminals. Always important, time and cost efficiency of port calls became critical. A large containership cost $40,000–60,000 a day to operate at sea and perhaps half that in port. Carriers sought not only ports whose location provided sufficient cargo to make the call worthwhile, but also marine terminals that minimized the carriers' costs through both competitive rates and efficient operations.

Competition Intensifies

Interport competition was stimulated by increasing awareness of the impact that a thriving port had on employment, incomes, tax revenues, and business location decisions. This awareness was heightened by a spate of port economic impact studies.[20] The potential effects of a port's growth—or decline—on the economic development of its region sometimes convinced governmental authorities to make or permit capital investments in port facilities that might not have been justifiable strictly on a basis of the investment's profitability or of national or regional capacity requirements. Some ports subsidized carriers' or shippers' operating costs in order to win their business. Overall, cargo handling capacity often grew faster than total available cargo in the surrounding area. Thus interport competition was very nearly a zero sum game, a struggle to shift the existing cargo base from one port to another. Rotterdam's gain almost surely would be a loss for Antwerp.[21]

Furthermore, gains and losses often occurred in big chunks. A ship line's decision to leave one port for another could mean a swing of half a million tons of cargo a year and with it perhaps 2,100 jobs and $36 million a year in revenue for the local economy.[22] The publicity surrounding the event often left the perception that one port city was growing and the other was dying—images that could become self-fulfilling prophecies.

With deregulated rail rates and the advent of regular double-stack rail services (see below), ports found themselves competing more intensively, not only against nearby rivals, but also against ports hundred or even thousands of miles away. In the late 1980s, most of the containerized cargo from Asia bound for some eastern U.S. ports did not arrive by ship but rather was discharged on the West Coast and landbridged by rail across the continent. The customary concept that each port had its own "captive" hinterland was eroded considerably.

Traditional Considerations

In this new environment some traditional considerations took on heightened significance. First, speedy access to major shipping lanes became even more important. If a big containership could save half a day's sail—and $20,000 to $30,000—by calling at Hampton Roads at the mouth of Chesapeake Bay rather than go up the bay to Baltimore, the Virginia port had a competitive advantage.

Second, being the first or last port of call on one side of an ocean had always been desirable. Now, it could mean selection by a major carrier as the most logical site for load centering or landbridging.

Third, although "captive" hinterlands had diminished, a port that was surrounded by a populous, affluent, and industrialized region still was strongly positioned. What big ship line operating in the North Atlantic could afford to ignore the "natural" cargo base of a port like New York, despite the congestion and expense of operating there? And once there, why not use its extensive road and rail connections to marshall containers from farther West?

Finally, sufficiently wide and deep channels, whether natural or manmade, became more important as containerships grew in width and draft.

New Investment

Traditional advantages were not enough, however. In order to compete for container business in the late 1980s, ports and terminals invested heavily in new port facilities, equipment, and information systems; provided a variety of nontraditional services; spent unprecedented amounts on advertising, marketing, and promotion; and carried on a complex set of relationships with politicians, government agencies, environmental groups, and other "neighbors." As the work changed, so, gradually, did the workforce.

Investments typically included deepened channels and berths as well as wharf lengthening to accommodate the larger "fourth-generation" containerships. Some ports built entirely new terminals designed exclusively for container operations, with acres of open space, well away from the congestion of the central city. Elsewhere, new gates and access roads were built, providing truckers with faster connections to interstates and auto routes. A number of ports added on-terminal rail facilities that expedited the operation of unit trains and raised the possibility of moving whole trainloads of containers directly between ship and rail.

The carriers' demands for faster, high-volume port calls and the increased ship sizes led to the development of larger dockside cranes. Using microprocessors and automation technologies, some of them could lift nearly one container per minute under ideal conditions—nearly double the rate of their predecessors. To support the new cranes, yard vehicles, and other equipment capable of moving several containers at once were developed.

Ports and terminals also invested heavily in the information technology described earlier. With intensified interport competition, "customer service" became an important priority, and that meant handling more efficiently information about the customers' containers as well as the containers themselves. Thus the use of computers was expanded rapidly to record all sorts of information about incoming containers and to assist decision making about terminal operations. Computers were also employed for electronic data interchange (EDI)—to communicate or exchange documents with shippers, consignees, agents, financial institutions, freight forwarders, customs brokers, and customs services as well as ocean and inland carriers.

Port investment capital was generated from a variety of sources. In the United States the most typical were revenue bonds, retained earnings and general obligation bonds, followed by local or state government appropriations and federal expenditures. However, a number of ports had their own limited taxing authority, revenue from other enterprises—airports, bridges, and tunnels, and rental income from retail or office space in marine terminals or elsewhere that was leased for nonshipping activities.

In the United States $3 billion was invested in port construction in the 10 years from 1973 to 1982, about one-third of it in container facilities.[23] The rate of investment then accelerated, and by 1988 16 of the biggest U.S. ports had $3.7 billion in construction projects programed for the next five years, raising new concerns about overcapacity.[24]

Nontraditional Activities

By the mid-1980s it was clear that the port facilities themselves were only part of a landside transportation system that had to function in an efficient and integrated manner for a container port to succeed. Traditionally, a port's supporting infrastructure had meant the roads and rail lines leading to and from its hinterland. Their development and management usually were someone else's responsibility.

Now, however, in order to compete more effectively, port authorities and terminal operators became involved in a variety of nontraditional activities. Some ports or terminals ran regular inland or coastal feeder barge services. Some developed inland rail collection points intended to bring the port "closer" to the shipper and consignee. Some organized and promoted unit trains (often using double-stack railcars) to provide fast direct service to and from the container terminal. In hope of generating additional cargo, a few incorporated their own export trading companies or obtained their own containers and functioned as freight consolidators.

Problems with Neighbors

Although port authorities and marine terminals long have had to coexist with others along the waterfront, in the 1980s the conflicting interests and concerns of their "neighbors" became matters of increasing sensitivity. Waterfront property became highly valued for residential, retail, or office uses. These uses were often cleaner and offered greater economic benefits to the immediate community than a container terminal. However, they also were likely to deprive the port of space needed for future expansion.

In some ports increased volumes of container traffic through residential and commercial areas created congestion, noise, air pollution, and road damage. Conversely, population and vehicle traffic growth posed operational problems for terminals and the truck and rail companies that served them.[25] Major facility construction projects stimulated environmental concerns about wildlife, fisheries, dredge spoils, and adjoining residential communities. Ports and terminals found themselves increasingly needing all sorts of specialized expertise with which to respond and adapt to these concerns.

The Marketing Mindset

All of these investments and activities were often carried forward with a strong marketing mindset. Heretofore, advertising services in trade publications and making sales calls on shippers were predominantly functions of the carriers themselves. Now, port authorities engaged extensively in both, as well as sending representatives to trade shows all over the world, publishing their own glossy periodicals and undertaking a variety of other promotional activities. Sometimes it seemed hard to be certain whether promotional activities were being devised to capitalize on the new investment or service—or vice versa.

"The cargo drives the calls" has long been a maritime maxim. It means that cargo tends to flow to the nearest port, and vessel owners send their ships to where the most cargo is. In the 1980s, however, the relationship between the cargo and the calls became more interactive. If, for example, a number of big carriers chose the same port as their load center and made arrangements with inland carriers for inexpensive, efficient door-to-door service, the situation could be reversed; the calls could drive the cargo. If this occurred, the port could spread its capital and fixed operating costs over more carriers. By doing this, it could keep its charges to them lower, attracting yet more calls and cargo. Under such circumstances, it was only logical for ports to vigorously market themselves to both shippers and carriers.

New People

The variety of new activities in which port authorities and marine terminals had become involved meant the hiring of new staff—data processing personnel, market and operations analysts, salespeople and public relations representatives, attorneys, and lobbyists. Similarly, as the nature of the most crucial survival needs changed, different sorts of people moved into positions of leadership. Senior executives came increasingly from marketing, financial, legal, and economic development backgrounds, and they often had extensive experience outside the shipping business.

INTERMODALISM MATURES

Perhaps the most profound effect of the environmental changes of the 1980s was the accelerated movement toward intermodalism. Visionaries like McLean had, of course, foreseen the potentialities decades earlier. However, the prerequisites to full development—changes in global commercial patterns and attitudes, development of new information and cargo moving technologies, and relaxation of government constraints—were not in place until the mid-1980s. Furthermore, it took some of the players several years more to recognize and utilize them.

The Railroads

For inland carriers, especially the railroads, big modifications were necessary. One was price competition. Particularly where large amounts of cargo were involved, railroads now competed on price, rather than simply offering standard common carrier tariff rates.

A second change was new capital and equipment investments—intermodal container freight stations for consolidating cargoes and shifting containers quickly from rail cars to trucks chassis; new yards nearer marine terminals; double-stack rail cars that carried more containers and made moving containers 30–40 percent less expensive than old-fashioned flat cars; rebuilt bridges and tunnels to accommodate the new cars; and all sorts of data processing and communications hardware.

Third, railroaders had to change their mindset, from simply offering rail transport to thinking in terms of integrated systems that provided "customer service" in the movement of international shipments. This meant developing new equipment such as "road-railers," which could run like a truck trailer on the road and like a rail car on tracks. It also meant negotiating collaborative relationships with truckers, ports, and ship lines, as well as sharing information previously closely held—about scheduling, cargo location and status, arrival times and the like—with these new "partners" and with customers.

Blurring Roles and Identities

Part of the intermodal transformation has been a blurring of organizational identities and roles. In the world of double-stack unit trains, who operates and markets rail services? The railroad? The port authority? The ship line? If ports, ship lines, and even railroads begin to conduct the functions of cargo expediters and the biggest shippers internalize those functions, what happens to the freight forwarders, customs brokers, and freight consolidators? Who does what is not clear. However, at least one answer seems to be the emergence of very large, integrated transportation conglomerates that offer comprehensive door-to-door service and that own or have subcontractors to provide every mode of transport as well as intermediary services. Among U.S. firms, CSX-Sealand and American President Company illustrate this prototype.

Another aspect of intermodalism is increasing competition among modes. A customary view of mode choices perceived a hierarchy of options, each of which involved a different mix of cost and speed as well as distance. This shipper's options began with the ship (cheapest, slowest) and moved through the barge, train, and truck to airplane (fastest, most expensive). By the late 1980s, however, the economy of double-stack cars and the reliability and speed of unit trains made railroads more competitive with both trucks and ocean carriers in some markets. Similarly, the struggle for certain types of cargo between ocean and air carriers intensified as competition within the air freight industry affected pricing, and some ocean carriers offered high-speed refrigerator service for perishable cargo.

Finally, there has been a gradual blurring of the distinction between international and domestic shipments. As domestic use of ocean containers increased and international cargo flows suffered imbalances, a few transportation companies recognized synergies between the two. In the mid-1980s, when eastbound imports from Asia filled the land-bridging unit trains, some operators, desperate to fill the westbound backhauls, sought domestic cargo moving westward from the East or Midwest.

Notes to Chapter 1

1. L. Royse, "Amphorae: Containerization of the Past," *The Mariners' Museum Journal* 18, no. 2 (Summer 1989):25.

2. R. S. Agman, Lectures on the maritime transportation industry, delivered at the College of Business and Public Administration, Old Dominion University, Norfolk, Virginia, April 4, 1988.

3. *Ibid.*

4. H. Takashashi, "Miserable Decline in Free World's Liner Shipping—Something is Wrong, Chapter 3," *Shipping and Trade News* (Tokyo), May 13, 1988.

5. Organization for Economic Cooperation and Development (OECD), *Maritime Transport: 1981* (Paris: OECD, 1982).

6. Organization for Economic Cooperation and Development (OECD), "Economic Outlook, No. 31," (Paris: OECD, July 1982); D. A. Ball and W. H. McCulloch, Jr., *International Business: Introduction and Essentials, 3rd ed.* (Plano, TX: Business Publications, 1988), Chap. Two.

7. W. Armbruster, "Export Surge Welcomed," *The Journal of Commerce*, June 20, 1988; R. Horowitz, "Surge in U.S. Exports Fills Westbound Ships," *The Journal of Commerce*, May 4, 1988; H. Takahashi, "Miserable Decline in Free World's Liner Shipping—Something is Wrong, Chapter 1," *Shipping and Trade News* (Tokyo), May 11, 1988.

8. Organization for Economic Cooperation and Development (OECD), *Maritime Transport: 1986* (Paris: OECD, 1987), pp. 19–23.

9. M. L. Chadwin, ed., *Proceedings: The Shipping Act of 1984: Evaluating Its Impact: A Conference Sponsored by the Federal Maritime Commission and Old Dominion University, Norfolk, Virginia, June 12–13, 1986* (Norfolk: Virginia Center for World Trade, 1986). Although deregulation and reductions in construction and operating subsidies programs in the United States constituted important changes for transportation service providers, other aspects of government policy in the U.S. and elsewhere, such as cabotage and cargo preference laws, changed little if at all.

10. W. Armbruster, "Electronic Data Interchange Expert Urges Shipping to Expand EDI Role," *The Journal of Commerce*, June 1, 1988.

11. For a brief account of that struggle, see M. L. Chadwin, "From Wharf Rat to Lord of the Docks: The Longshoreman—Yesterday and Today," *Oceanus* 32, no. 3 (1989). For a more detailed version, see M. Russell, *Men Along the Shore: The I.L.A. and its History* (New York: Brussel and Brussel, 1966).

12. J. Nolan, "NY Roster of Dockers Shrinking, *The Journal of Commerce*, May 26, 1987.

13. B. Barnard, "U.K. Ports Blast Dock Labor Laws," *The Journal of Commerce*, March 13, 1986.

14. M. Magnier, "Six-Man Ship Crew Wave of the Future?" *The Journal of Commerce*, October 16, 1986; J. Porter, "Swedish Shipowners to Try Smaller Crews," *The Journal of Commerce*, August 16, 1988.

15. OECD, *Maritime Transport*, pp. 64–76.

16. M. Magnier, "Radical Plan Offered to Cut Overcapacity," *The Journal of Commerce*, October 4, 1986; OECD, *Maritime Transport*.

17. Chadwin, *Proceedings*; B. Mongelluzzo, "Shippers, Carriers Hold Different Views on Service Contracts," *The Journal of Commerce*, February 22, 1988.

18. "Landbridging" means a cargo movement that crosses a body of land between two ocean legs (for example, eastward by ship across the Pacific to the West Coast of the United States, by train to a U.S. East Coast port, and across the Atlantic by ship to Europe). "Mini-landbridg-

ing" or "minibridging" means a movement that crosses one ocean by ship and then crosses a body of land but ends at a seaport on another ocean. "Microbridging" means a movement that crosses one ocean by ship and then proceeds by rail to an inland location.

19. OECD, *Maritime Transport*; B. Barnard, "Analyst: Jumbo Ships Will Soon Dominate," *The Journal of Commerce*, December 2, 1987.

20. U.S. Maritime Administration, *The Regional Port Impact Model Handbook, Volume I: Guide for Preparing Economic Impact Assessments Using Input-Output Analysis*, prepared by the Planning and Development Department, Port Authority of New York and New Jersey (Washington, DC: U.S. Department of Transportation, Summer 1982); G. R. Yochum and V. B. Agarwal, *The Economic Impact of Virginia's Ports on the Commonwealth: 1984* (Norfolk: Old Dominion University Maritime Trade and Transport, 1986).

21. For example, see "Sealand's Contract Breathes New Life into ECT," *The Journal of Commerce*, February 4, 1988.

22. See Yochum and Agarwal, *Economic Impact*, for the calculation of values of a ton of containerized cargo in terms of employment, incomes and tax revenues in the Port of Hampton Roads.

23. U.S. Maritime Administration, *Public Port Financing in the United States, Volume One: Executive Summary* (Washington, DC: U.S. Department of Transportation, June 1985).

24. G. Joseph, "U.S. Port Investment Will Rise Sharply; Overcapacity Feared," *The Journal of Commerce*, January 11, 1988.

25. For an analytic approach to these problems that utilizes computer simulation, see J. A. Pope, T. R. Rakes, and L. P. Rees, *The Interaction of Road Traffic Flows in the Port of Hampton Roads: A Research Report* (Norfolk: Virginia Center for World Trade, 1988).

Chapter 2
THE MARINE CONTAINER TERMINAL AND ITS OPERATION

The marine container terminal is the place where containers received from ocean vessels are transferred to inland carriers, such as trucks, trains, or canal barges—and visa versa. Thus it is a major node in any intermodal transportation network.

At first glance the operation of a container terminal is deceptively simple. In fact, it closely resembles a manufacturing "job shop." Containers arriving at the terminal are the "raw materials." Although most containers undergo no physical transformation, terminal workers and equipment perform various operations on the containers, which give them added value. These operations may include unloading the containers from the vessel, staging them for speedy access by the next carrier, or mounting them on "chassis," specially designed truck trailers that have mounting pegs and locks on each corner to hold the containers fast. Eventually, the containers leave the terminal as "finished products."

The marine terminal must perform many of the same activities as any manufacturing facility—inventory control including raw materials (incoming containers), work in progress (stack and chassis management), and finished goods (staging and clearing containers); controlling the work performed through work orders; establishing priorities for operations; maximizing productive time by minimizing set-up activities; and maintaining cost accounting procedures that provide a basis for pricing, billing, and allocating resources.

In the typical job shop, however, raw materials or parts of the same type are interchangeable. If a part is needed, a worker goes to a bin containing that type and takes any one he finds. In the marine terminal, however, when a trucker calls to pick up an inbound container (an "import box" in the jargon of the industry), he does not want just any import box. He is responsible for picking up and delivering one particular container, the one that holds his customer's cargo.

Similarly, when a vessel is being loaded with "export boxes," only a few hundred specific containers among the thousands on the terminal are supposed to go on board. Through careful "stowage planning," the containers going aboard must be placed on the vessel very quickly in locations that permit them to be discharged efficiently later. If not, time will be wasted—during both this port call and ensuing ones. Each minute

19

that a large containership sits at a terminal being loaded and unloaded may cost the ship line $100 or more. Thus efficient loading and discharging are critical to profitability.

These distinctive requirements have some important implications. First, an accurate and timely flow of information about the containers is, arguably, as crucial to efficient terminal operation as the careful handling of the containers themselves. Thus this chapter describes not only the cargo-handling activities and equipment that may be found in a container terminal, but also management information systems (MIS) used in container management. We look at alternative forms of MIS and how the participants at the marine terminal interact with each other through the MIS.

Second, if we resume the job shop analogy, it becomes clear that the most convenient way to run a marine terminal is to keep each container on a chassis, a so-called all-wheel operation. Material handling is minimized, since over-the-road truckers take the container and its chassis to or from its "place of rest" on the terminal, and a terminal worker or stevedore can later hook a small truckhead called a "yard hustler" to the chassis and move the container quickly to the side of the ship. Containers may be retrieved in any sequence, since all are equally accessible. In job shop terminology, an all-wheel terminal does not require long production runs.

On the other hand, containers that are stacked on top of each other upon arrival in the yard occupy much less space. However, they are more difficult to retrieve and move, and their retrieval requires specialized yard equipment such as transtainers, gantry cranes that ride back and forth over the stack of containers on rubber tires or rails and have "spreader bars" that lock into the corner fittings on the top of a container. Operating a container stack efficiently implies advanced knowledge of the voyage (vessel, departure date) and destination of each container in the stack so that work can be sequenced to minimize transtainer set-up times.

The advantages of an all-wheel operation are gained, however, only by incurring costs similar to those in a manufacturing situation. Gaining the advantages of short set-up times and minimal handling in a factory generally requires substantial capital investment. Similarly, a marine terminal that maintains its entire inventory of containers on wheels must invest in more land to accommodate the wheeled operation. Frequently, however, the terminal is in a location where acquisition of adjoining land is either very expensive or impossible in the short term. Furthermore, although the terminal itself may need to invest less in yard equipment such as transtainers, someone—usually the steamship or truck lines—will need to invest in more chassis.

Earlier we made several references to efficiency in terminal operations or container handling, and we do so repeatedly in the following pages. The term "efficient container management" can be defined in a number of ways, but in general it refers to three objectives. The first is minimizing the time the container is on the terminal. A container sitting at the terminal is costing the shipper money and utilizing valuable space.

The second objective is minimizing the number of times the container is moved while on the terminal. Each move ties up equipment and labor and generates costs that must be borne by the terminal, the ship line, the shipper, or some combination of them. The last objective is to minimize the time each containership remains in port, since a ship is most productive when it is moving. Sometimes one of these objectives may be violated to improve the attainment of another. For example, a container may be moved from its initial place of rest to a staging area to decrease the time necessary to load it on the ship.

AN OVERVIEW OF TERMINAL OPERATIONS

Terminal Functions

Every marine container terminal performs four basic functions: (1) receiving, (2) storage, (3) staging, and (4) loading (see fig. 2.1). All four functions must be performed for all containers, whether they are imports (and thus enter the terminal from a ship and usually leave by land) or exports (and thus usually enter the terminal by land and leave by ship).

Receiving is the function of having a container arrive at the terminal, either as an import or export, recording its arrival, and capturing the relevant information about the container. Storage is the function of placing the container on the terminal in a known and recorded location so it may be retrieved when it is needed. Staging is the function of preparing a container to leave the terminal. For example, an export container may be staged at the time of initial storage, or it may be moved from its initial place of rest to a second location with other containers for the same ship. The staging function

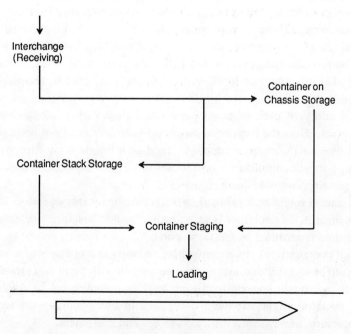

Figure 2.1. A container terminal layout showing different functions.

(sometimes called "prestowing") requires that information about the container be matched with the manifest for the ship; in other words, the containers that are to be exported are identified and organized so as to optimize the loading process. Import containers have similar requirements, except fewer containers will generally move out of the terminal together. Finally, the loading function involves placing the correct container on the ship, the truck, or other means of transportation. This function is especially critical for export containers, since the wrong container on a ship may travel long distances and may not be unloaded until the ship is stripped of all containers.

In addition to the above four basic activities, other functions may take place on the terminal. For example, the container and chassis may be inspected for damage before entering or before leaving the terminal. Containers may be inspected by customs officers or by other government authorities such as food inspectors or security officials. Furthermore, some containers may be packed or unpacked ("stuffed" or "stripped" in industry parlance) at the terminal. This is because some containerized cargo involves shipments that are "less than container load" (LCL) in size. These small shipments, often from a number of different shippers to a variety of consignees, must be "consolidated" into a single container, either at a marine terminal or an inland facility, such as a warehouse or container freight station (CFS).

The Participants

Different participants interact with each other at each of the container terminal functions. The principal participants fall into six categories: (1) the shipper who loads the container and sends it to the terminal, (2) the inland carrier who transports the export container from its origin to the terminal and the import container from the terminal to its inland destination, (3) the terminal operator who managed "interchange" (the landside entry and exit of containers), container control and yard operations, and use of the terminal's wharf space and equipment, (4) the stevedore who loads and unloads the containership, (5) the containership line, and (6) the consignee or recipient of import cargo.

The exact identity of each of these participants may vary, although the functions themselves do not. Thus the shipper of an export container need not necessarily be the producer or owner of the cargo. Instead, it could be a freight forwarder serving as the owner's agent, a freight consolidator, or a broker of carrying services such as an NVOCC (nonvessel operating common carrier).

The inland carrier might be a railroad, a truck line, or a barge operator. It also might be a drayage firm, if the container is to be hauled a short distance (for example, to or from another marine terminal in the same port).

The terminal operator might by a public port authority that operates a "common user" terminal open to any vessel that makes arrangements to call there or a stevedoring firm operating such a terminal under a long-term lease or management contract. However, the terminal operator might also be a containership line operating the terminal as a "dedicated" facility, serving only its own vessels and customers.

Similarly, the stevedoring function is organized differently in different terminals. Frequently, it may be performed by the terminal operator and included as part of yard management. However, in many U.S. terminals the stevedore is an independent contractor hired by the ocean carrier. For export containers, the stevedore is responsible for removing the container from the storage or staging area and loading it on the ship. For import containers, the stevedore removes the container from the ship and places it in storage. In some terminals, the stevedore does stowage planning for the ship, stages containers before the vessel arrives, lashes above-deck containers to the ship, and assists in linehandling when the ship ties up and casts off.

At smaller ports the ship line may have no employees of its own but may engage a ship's agents to make all necessary arrangements with the terminal and stevedore. In addition, the ship's agent may be responsible for paying all charges incurred by the ship, looking after the needs of the ship's company, and handling relations with shippers, consignees, and government officials.

Finally, the consignee of an import box might not be the retailer who bought the cargo. Instead, it might be a customs broker acting as the retailer's agent, who insures that all fees on the containers are paid and that "releases" permitting its prompt departure from the terminal are secured from the ship line, customs, and the terminal itself.

Terminal Equipment and Organization

As the prior section explained, there are a number of ways of organizing a container terminal. Each can impose different information flows and data needs. Since the common user or public terminal, which serves many shippers and ship lines, poses greater operational and information management problems than the dedicated terminal serving a single line, we concentrate here on the public terminal.

Whether public or private, all true container terminals have several things in common. The first is that they have invested in one or more huge ship-to-shore gantry cranes. Typically, these cranes run along tracks that adjoin a long length of wharf. They can be moved along the track so that several can be "ganged" to work on a large vessel simultaneously. Usually, 800 or 900 feet along the wharf is allocated to a "berth," but, since the wharf is lineal, vessels can be accommodated wherever space permits.

A typical ship-to-shore crane can lift 30 to 50 tons and load or discharge between 25 and 50 containers per hour, depending on its age and design. By 1990 the newest and most sophisticated ones cost about $5 million. They were extensively automated and utilized two "trolleys" so that they processed two containers at once. Furthermore, they could reach out across 16 rows of containers on board a ship, thus accommodating the larger-than-Panamax vessels that began entering the global container fleet at the end of the 1980s.

A second characteristic of a true container terminal is that the area immediately behind the cranes is devoid of storage sheds, warehouses, and other buildings for a distance of at least several hundred feet. This permits the staging and rapid movement of containers without obstruction, as well as the stacking of containers close enough to the wharf to minimize cycle times between the ship and the storage pad. The area usually is constructed of reinforced concrete to support fully loaded containers and heavy yard equipment. The only structures visible are light posts to permit night work, temporary barriers to guide critical traffic flows, and perhaps a control tower manned by yard supervisors.

A third and defining characteristic is, of course, the presence of hundreds of ocean containers. As indicated in Chapter 1, the majority currently in use are eight feet wide, eight and a half feet tall, and either 20 or 40 feet long. However, many variations exist. Thus there are so-called high cubes that are nine and a half feet tall and offer the shipper more cargo space. Similarly, some ship lines or container leasors offer nonstandard lengths, such as 45 or 48 feet. Although most dry cargo containers are made of either steel or aluminum, other materials are sometimes used. Differences in design and reinforcement also yield different carrying capacities. Typically, however, most 20-footers can accommodate about 10 tons, and most 40-footers about 20.

The special needs of shippers also have led to the construction of refrigerated containers (both electrical and chemical), ventilated containers, open tops, half-heights, side openers, tank containers, flat-racks (which have neither sides nor tops), and "Sea-Sheds" (that may be three times as wide and tall as a standard container to accommodate oversized construction or military equipment). Regardless of their design, most containers carry a unique ISO (International Standards Organization) identifier usually consisting of four letters designating the owner of the container and four numbers identifying the specific box itself.

As noted earlier, terminals do differ in how they store and move the containers within the yard. The most prominent "pure" types of terminal storage organization are chassis storage, stack-with-transtainer storage, and stack-with-straddle carrier storage. In chassis storage, the container remains on the over-the-road truck chassis and is not removed from the chassis until it is loaded on the ship. Transtainer storage involves moving the container into and out of the stack by transtainer. The container must be transported to and from the transtainer using some sort of chassis. A straddle carrier can both transport the container from place to place and stack the containers. Other equipment that can be used to both stack and transport includes side loaders (heavy duty fork lift trucks), top loaders (fork lift trucks equipped with spreader bars), and reach stackers (essentially small mobile cranes). (See Fig. 2.2 for illustrations of typical terminal equipment.) More fully integrated systems that provide continuous flows of containers between the ship and the storage area via conveyors have been designed, but only one (Matson Line) is reportedly operational at this writing.

Most terminals, in fact, have a mixture of the different types of storage organization. Some may use chassis storage as a buffer for the stack operation, but others may move containers directly from chassis to ship. The particular variation also may depend upon the desires or requirements of the shipper or the shipping line. The amount of land available may also determine the variation. Straddle carriers tend to be more flexible and mobile than transtainers but require more land, since a straddle carrier stack must be only one container wide and is rarely more than two high. A transtainer operation, on the other hand, may stack five to seven containers across and three or four high.

Terminals also may differ in how they receive containers from inland. Virtually all terminals receive containers over the road by truck. Most large terminals also receive substantial numbers of containers by rail. Rail receipts may be directly to the terminal or to a switching yard from which the containers are drayed into the terminal. The terminal also may receive containers by barge, for example, via an intra-coastal waterway, as along the East Coast of the United States, or through canals such as in Europe.

Rail and barge arrivals differ fundamentally from over-the-road deliveries in the number of containers involved. Each truck brings in only one or two containers; each barge or train involves the arrival of 100 or more containers at once. Usually, barges are worked on the same wharf as containerships, although some recently constructed terminals, such as European Container Terminus' (ECT) Maasvlatke terminal,[1] have separate slips and cranes for loading and discharging barges.

Most on-terminal rail deliveries are received in an area of the terminal close to but separate from the wharf and the main storage areas. Direct rail-to-ship transfers have been much discussed in the industry, since they would eliminate the need to shuttle containers between the rail siding and the ship-to-shore crane or staging area. However, coordinating the stowage sequence and arrival times of specific unit trains with those of specific vessel calls poses a substantial obstacle, and we know of no terminal where direct rail-to-ship transfers is the practice at present.

Finally, the amount of land available may determine the technology used or how a terminal is laid out. As noted earlier, a chassis-oriented terminal requires more space than a stack-oriented one, and a straddle carrier operation requires more land than one that uses transtainers. If sufficient land is available, the functions of receipt, storage, staging, and loading may all be in one contiguous area. It may be, however, that the functions must be separated into different geographic areas, imposing additional burdens on the MIS. For example, some terminals in Europe and the United States encourage the storage of empty containers or chassis in separate facilities (sometimes called "drop-lots" or "container parks") away from the shore-side terminal. A few terminals in East Asian ports, where space is at a premium, use less costly inland locations for staging and only begin transferring export containers to the main terminal when the vessel is ready to be loaded.

In addition, several ports in the United States (in North Carolina and Virginia) and Europe (ECT-Rotterdam) have developed "inland terminals" more because of competitive marketing strategies than space constraints. By receiving containers several hundred miles inland and putting them on unit trains to the port, a terminal can reduce shippers' costs and, in effect, deepen its hinterland. Regardless of the rationale, dispersed facilities require the networking of equipment and information and the development of information systems able to meet the needs of a more complex set of users.

In the next section, we sketch out some of the general requirements for an MIS at a container terminal rather than entering into a detailed discussion of the requirements for every combination of variation. Our purpose is to explain the information flow

Figure 2.2. (a) Ship-to-shore crane. (b) Transtainer (yard gantry crane). (c) Straddle carrier. (d) Top loader. (*Courtesy*: Virginia Port Authority)

function and its importance for meaningful decision making by the managers of the marine container terminal.

AN OVERVIEW OF MANAGEMENT INFORMATION SYSTEMS

In order to support meaningful decision making in any setting, an MIS must generate information with certain characteristics. The two most pertinent characteristics are accuracy and applicability. Without these two, an MIS may generate an enormous amount of information, but little of it would be useful for the decision maker.

Accuracy is a relatively simple concept. The data that are entered into the MIS must be correct when entered, and any processing or transformations of the data must be done correctly. This implies that the MIS must not only be structured to allow operators to enter data correctly, it also must be structured to assist and encourage them to enter data accurately. Conversely, the MIS must not be structured so that errors are made easily and without the ability to identify the source of the errors.

Applicability has several dimensions. In general, it means that the information generated by the MIS must be relevant to the decision at hand. In addition, it means that the information required by the decision maker must be complete. Information can fail to be applicable in several ways. One way is simply to have nothing to do with the decision. For example, the age of the over-the-road driver has nothing to do with the decision as to where to place the container after it passes through interchange. The information is irrelevant and inapplicable no matter how accurate it happens to be.

Another way for information to be inapplicable is for it to lack timeliness. In other words, information fails the applicability test if it is the right information at the wrong time. An example might be the data on a shipment of containers arriving at the terminal several hours after the containers themselves.

A final way information may fail the test of applicability is to be in the wrong place. At a container terminal, if the MIS personnel have all the accurate and timely information necessary for placing a particular container, but the information is not available to the person who must make the decision, the information fails the test. Any analysis of the information flows in the MIS at a terminal must be done in the framework of evaluating their accuracy and applicability.

Two terms in computer jargon that are related to the concept of applicability are "real time" and "on-line." A real time system is one in which the data retrieved from the system are current. If there is a delay between when information is collected and when it is available to users, then the system is not real time. Although information does not have to flow from a real time system to be applicable, one must always consider carefully the trade-off between the cost of a real time system and the potential for loss of applicability. An on-line system allows users to query the data base directly for information instead of requesting the information when it is needed or receiving information through regular reports. In a system that is not on-line, the data may suffer a loss of timeliness because of the time lag between the request for and the receipt of the information. An on-line system, on the other hand, does not necessarily contain the most current information; in other words, it is not necessarily real time.

TERMINAL INFORMATION SYSTEMS

The basic decisions that must be made by operating personnel at any container terminal are where to place containers and/or chassis arriving at the terminal and where to find containers and/or chassis leaving the terminal. This imposes two requirements on the MIS: (1) it must keep track of where units are located and not located, and (2) it must provide information about the relevant variables to be used in the yard management decisions. The first requirement may be fulfilled by simply recording a unique coordinate (or location) on the terminal for each unit. The second requirement implies the provision of additional characteristics about units such as their ship line, if they are export or import, and if they are in special storage locations, such as those for refrigerated containers or containers holding hazardous cargo.

Although the management information systems at modern container terminals are ostensibly computer-based, many informal manual "data bases" are often used. Of course, manual "data bases" tend to be employed at early stages of container terminal development, but they also spring up later to compensate for times when the computerized data base is unable to provide this accurate and applicable data for the decision makers. An example would be a yard foreman's hand-drawn sketch of yard organization showing container locations. The sketch is cumbersome to generate and update, but the information is necessary for yard management. Foremen would use such a sketch if the MIS did not provide the same information accurately and at the right time and place. An MIS that provides accurate and applicable information will remove the need for the manual systems and consequently cause them to disappear.

Let us now follow the flow of information through a container terminal to pinpoint where the critical points are for accuracy and applicability and to describe systems for dealing with them. (See Fig. 2.3 for an illustration of some of the most important flows.)

Collecting Interchange Data

In the receiving function, the information on a container usually originates with the shipper for export containers or the shipping line for import containers. The shipper must inform the terminal which container is arriving, what it contains, where it is going, and on which ship it will be transported (among other things). The terminal receives the information at a container control office.

Containers arriving over the road first pass through an interchange system. At this point data concerning the container—identification number (ID), the chassis ID, the contents, the weight, condition of the box and chassis, and destination—are captured.

The data would appear to be timely, since at most terminals the data are entered

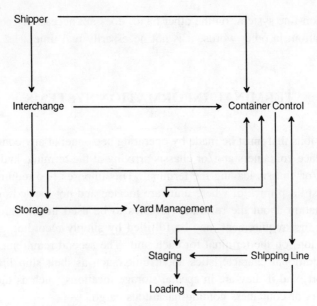

Figure 2.3. Container terminal information flows.

directly into a computer system for container management. Some terminals have found that even this is not timely enough. If a terminal guarantees that a container arriving within 20 minutes of departure will be on the ship, the terminal must have all the data ahead of time and simply confirm it when the container enters the terminal. If the terminal has enough land available, they may place a "preinterchange" or inspection station a mile or more before the primary interchange. The data on the container are forwarded to container control to allow time to check the container data while the driver is enroute to the primary interchange. This means that part of the interchange function is taking place while the container is enroute to the terminal, and thus the total time the container must wait for clearance is shortened. Accuracy is critical at this point, since a container could be temporarily lost if it is not correctly identified at the interchange function.

In an effort to speed the process even further, some terminals have modified the interchange process by reversing the information flow as much as possible. As advance information about export container arrivals becomes available (either in hard copy form or by electronic transmission), it is entered into the system by container control. As each container arrives at the interchange, the container number is transmitted from the interchange writer to someone in container control. If the container information has been previously entered, a form is either printed out at container control and sent to the interchange writer by tube, or printed out directly at the interchange booth. The interchange writer completes the form and sends it back. If the container has not been previously entered in the computer, the interchange writer would proceed to record the data about the container. Decreasing the number of times data on the container must be recorded increases the accuracy of the data in the system by eliminating possibilities of incorrect entry.

Terminal Yard Information

After leaving the interchange system, containers move into the yard management system (see Fig. 2.1). A typical yard management system in the United States initially leaves the containers and chassis in an assigned place; they are later transferred to the stack using terminal drivers. In this system, containers are sent to a location checker who then assigns a place to park the chassis and containers. The checker records the container ID and the parking location for later entry into the MIS. At this point, both accuracy and applicability of the data tend to break down. Drivers may drop their loads at the first available empty parking space instead of ever getting to the location checker. If they do proceed to the checker, he may record the container number and/or location incorrectly. If these are recorded incorrectly, the error may not be identified until the next day, unless the container is specifically called for and is the subject of a physical search. Even if errors are corrected within 24 hours, during that period the container and its chassis are effectively lost, and the cargo could miss its vessel. Although the data become accurate after being corrected, the timing of the correction may cause them to lose their applicability.

The crux of the problem in receiving a container into the yard is the correct recording of its location, or, conversely, the placing of the container in the correct location. In many terminals, the interchange and the yard are connected to the computerized container control system. When a container passes through the interchange, it shows up on a screen in the yard. A driver who attempts to bypass the yard system is noted within minutes. The assigned location is entered directly into the system (as at Hamburg's HHLA—Hamburger Hafen- und Lagerhaus-AG). In more advanced systems, the container location is assigned by a computer algorithm designed to minimize the movement necessary to ultimately move the container to the ship (as at Koper in Yugoslavia).

Checking to see that the container is actually placed in the correct location is more complex. Facilities that use transtainers may have detection devices on the equipment that tell the computer a container's location. Such systems tend to be difficult to calibrate and depend upon the operator entering the correct container number. Straddle carrier operations may have a system such as the carrier driver entering the number of the container in an adjacent location. The computer may then indicate an error if the locations are inconsistent. The error, however, may be in the original container's location, the check container's location, or the check container's number. The computer can only detect an error, not specify its nature or what is necessary to correct it. Detecting an error immediately, however, allows for its timely correction.

No matter how sophisticated a terminal's computerized system, it cannot be any more accurate than the people entering the original data. Ways of improving that accuracy are examined in Chapter 4.

Tracking Chassis

Terminals that are chassis-oriented face special problems. If the container is to be placed into the stack, the terminal need not take special note of the chassis on which it arrives.

If the chassis is to remain on the terminal, however, the information system must capture the data on the chassis. One weak point in some marine terminal computer systems is that a separate record is not created and maintained for the chassis as long as it is mated with a container. In such a case, a loaded chassis may be located only if the identity of its container is known. To minimize the problem, the system used to capture information about the container must be designed to allow creation of a separate record for the chassis.

The Stacks

In Chapter 4 we shall cover stack management in greater detail. Nevertheless, some points may be made about how computer data base accuracy and applicability can facilitate efficient stack management. In a stack-oriented system (either transtainer or straddle carrier) that has heavy utilization and throughput, it becomes critical to have the stacks properly organized and to have timely organization information available in the data base. We have addressed the problem of accurate and timely information on the location of chassis that are still attached to containers. In the stack, of course, the location information must be expressed in three dimensions—the row, the location in the row, and the height. Any one of the dimensions entered incorrectly could cause problems, although the proper row is most critical.

Staging

Regardless of the storage mode, information on the storage location must be matched with the container and placed in the container data base. This information must then be available to the yard managers and foremen to allow them to organize further storage and staging. For a container to be staged, its information must be matched with the requirements for an arriving ship. Whenever possible, staging is done in conformity to the stowage plan for the vessel so that loading can proceed at a high speed, without being interrupted by the need to search the chassis lot or stack for the "right" box to be loaded next.

Finally, those responsible for loading the ship must record the information that the container has, in fact, been loaded in a specific location in the vessel and is on its way.

Advance Information from Rail and Ship

At terminals with rail heads, there are additional data problems. The dominant information problem with arriving rail shipments is often the timeliness of the data. Except in the case of unit trains (those assembled at a distant point and transported unbroken to the terminal, often on behalf of a single containership line), it is sometimes difficult to get railroads to provide advanced information on containers enroute to the terminal. As a result, containers may arrive by rail unannounced or with no indication of their voyage, ship line, or destination port. The paperwork may follow the train much later, and phone confirmation of the destinations may not be possible immediately. The information will eventually be available and accurate, but its lack of timeliness causes it to fail the applicability test. The most sophisticated system for managing the stack cannot work if the information is not available when it is needed.

Import containers arriving on ships pose similar but less critical data accuracy problems. The problems are less critical for two reasons. First, most containership lines are conscientious about providing manifest data sufficiently in advance of the ship's arrival so that the data may be in the computer when the vessel berths. Second, import containers do not need to be managed as tightly as export containers, since it is the recipients' responsibility to pick them up rather than the terminal's to see that they are on the correct ship.

Nevertheless, frequently basic information such as the number of containers to be off-loaded is not available because of last-minute decisions by the containership line to load containers at the previous port. Often, but not always, the "surprise" containers are empty containers that tend to be managed differently from full containers (since the value of their cargo is zero). If the terminal has high land utilization, it could be in a position of having these "surprise" containers arrive without a place to put them. Even the most advanced terminal management system loses its effectiveness, if key data are not available when needed.

The industry has been addressing this problem by seeking to build connecting links between the terminal MIS and those of the ship lines and such inland carriers as railroads. Cargo information could be transmitted, exchanged, and confirmed 24 hours a day, if a suitable communications link among the computers were available. There are many difficulties in developing and installing such a comprehensive system. Protocols must be established among a large number of participants and often other projects are in place or in progress, such as an automated link with the customs service, which may use protocols unsuitable for such a comprehensive system. Further complications arise when protocols must be established among participants in different countries.

Terminals with a significant number of containers leaving by rail are an important exception to the statement that import containers' departure are primarily the recipient's responsibility. From the terminal's point of view, containers leaving on a unit train are similar to export containers leaving by ship. A large group of containers leave at once, and the terminal is responsible for assuring both that the correct containers are loaded and that they are loaded on the proper rail cars.

SUMMARY

This chapter describes the basic activities occurring at a marine container terminal (receiving, storage, staging, and loading) and the principal participants in those activities (the shipper, the inland carrier, the terminal operator, the stevedore, the ship line, and the consignee). Each activity and participant are present in all container terminals, although their identity and organization may vary.

Applicable and accurate data must pass among the participants for the efficient management of containers in a marine terminal. The use of a computerized MIS is virtually a requirement for such information but does not guarantee that it will be provided to the managers. This chapter describes the requirements for a computerized MIS and outlines some of the techniques for insuring that the information it provides is accurate and applicable. We also describe some of the problems in providing these data at each of the steps in processing a container. Chapter Four, which presents techniques for evaluating terminal performance, suggests possible solutions for these problems.

Notes to Chapter 2

1. In July 1989 ECT merged with two stevedoring firms that were extensively involved in breakbulk cargo, and the new entity was named European Combined Terminals.

Chapter 3

COSTING TERMINAL OPERATIONS AND MEASURING CAPACITY

The costs at a marine container terminal may be affected by the activities of six different decision makers: (1) the terminal operator, (2) ocean carriers, (3) inland carriers (usually truck or rail), (4) shippers, (5) harbor support auxiliaries (such as tugboat firms and pilotage organizations) and (6) government (local, state, and federal). Each of the six generate costs for themselves or for others. For example, the terminal operator absorbs expenses in providing berths and cranes; the ocean carrier must defray vessel and crewing costs while in port; the inland carrier bears costs related to vehicles and labor involved in moving cargo to and from the marine terminal; the shipper of containerized cargo bears inventory expenses, since his cargo is in-transit inventory; harbor support auxiliary firms have expenses related to the vessels, fuel, and labor they use in berthing and unberthing containerships; and governments incur costs in providing such services as harbor channels and navigational aids.[1]

In costing marine container terminal operations, it is important to consider the activities (and their costs) of all decision makers at a terminal. Specifically, the activity of one decision maker will invariably impact on the activities (and thus the costs) of the other decision makers at a terminal. For each decision maker to make proper decisions in the allocation of its resources at a marine container terminal, it should have an understanding of how its resource allocations impact on the activities and costs of other decision makers. For example, an ocean carrier considering initiating service at a marine container terminal might wish to evaluate the impact it would have on terminal activities and thus on the level of service it is likely to receive at that terminal. One impact would be on the number of ships arriving (the ship-arrival rate) at the terminal. An increased arrival rate would increase the likelihood of a ship having to wait for a berth at the terminal; this would increase the time and costs incurred by the ship in port. The increased arrival rate also would increase the demand for ship berthing assistance. The ship's cargo would have an impact on the terminal storage space required. Further, the increase in terminal cargo may cause inland carriers to have to wait at the inland gates of the terminal.

In addition to using tools for analyzing the cost of terminal operations, a marine terminal operator may also use terminal capacity measures in his planning process. The optimum capacity (or throughput) of a marine terminal based upon its current operations may be compared with its current (or predicted future) throughput by the terminal operator for deciding whether to expand terminal facilities.

The purpose of this chapter is to show how the terminal operator may develop and use computer-based analytic models to aid in decision making. In a time when containerships are becoming larger, ship operators are demanding faster turnaround times, and competition among ports throughout the world is becoming keener, terminal operators should be aware of the tools available to them. This chapter is not a complete "how-to" manual for constructing cost and capacity models. Its purpose is to show how cost models for the six decision makers at a marine container terminal may be developed and used to analyze major decisions. These models are intended to be representative, not definitive; they are not valid for every terminal in every port. Rather they are intended to show the factors that must be considered in constructing such models and how those factors relate to one another. Although the specifics for any given port may be quite different, the thought process required to construct the model should be similar.

Once constructed, the models can be applied in practice for determining (or predicting) the costs that will be incurred by each decision maker. Since these cost models consider interactive effects among the activities of the terminal decision makers, one may predict the effects on all the actors of decisions by one decision maker.

Following the sections presenting terminal cost models for each of the six decision makers, we discuss how these cost models might be included in a computer simulation. (More detail on the computer simulation technique is contained in Chapter 4.) Then, we discuss terminal capacity measures. Although the discussion is quite technical at times, wherever possible we have substituted clarity for technicality. Readers with quantitative backgrounds are referred to the notes for more complete coverage of the technical details.

COST MODELS FOR TERMINAL DECISION MAKERS

In the following discussion, we construct simplified models illustrating a port with a single container terminal handling one type of cargo and using one type of storage area (stack or chassis). Using a simplified model illustrates all the same principles and processes as a more complex model without getting into convoluted discussions and algebraic notation. One subscript that appears throughout the various models is the letter 't', which stands for a particular year. Many of the costs and activities must be looked at over a number of years, so the model must reflect which year is relevant.

Terminal Operator

The annual costs (C_t) associated with the operator of terminal facilities may be described as the sum of two components:

$$C_t = C_{th} + C_{ts} \tag{1}$$

The term C_{th} represents annual terminal freight handling costs. These costs include expenses related to: (1) terminal capital such as land, wharves, and interchange booths, (2) terminal equipment such as cranes and forklifts, and (3) labor, both direct (yard workers), and indirect (administrative personnel). The term C_{ts} represents annual terminal storage costs. These costs are expenses borne by the terminal operator and are related to the purchase, maintenance, and operation of terminal storage sheds, warehouses, and storage areas.

Freight handling costs. Terminal freight handling costs may be further subdivided into fixed and variable costs. Fixed costs include land, since the amount of land available to a terminal tends to be fixed over long periods of time. Determining the value associated with the land is difficult unless the terminal rents the land for a realistic fee. If the terminal or port authority owns the land, one must determine an economic rent reflecting alternative uses of the land. Using the approach recommended in the U.S. Maritime Administration's report, *Usage Pricing for Public Maritime Terminal Facilities*,[2] for example, the terminal or port authority must determine what a realistic rent would be if they did not own the land.

The purchase of capital equipment such as cranes, forklifts, vehicles, and mobile cranes is included in fixed costs because they normally must be budgeted for a year or more in advance by terminal operators and thus do not fluctuate with the short-run level of terminal activity. Hence, annual terminal fixed costs incurred in handling cargo would include both annual depreciation costs of capital and terminal indirect (or administrative) labor costs.

Variable costs pose another type of problem. The variable costs of handling cargo are often based on average throughput in these types of models. For example, total variable cost is the average number of containers per day times cost per container (times number of days per year for total annual cost). We use this approach here for simplicity, but a warning is in order. This approach ignores two facts—that (1) the amount of cargo passing through the terminal at any point in time may vary widely from day to day, and (2) it cannot exceed the capacity of the terminal to handle the cargo. If one uses an average throughput to do variable cost calculations, capacity is largely ignored since the average does not (and cannot) reflect the variations in throughput. So, rather than using average throughput, a better approach is to look at the amount of cargo taken on and discharged by individual ships that use the terminal. It requires more effort on the part of the model builder, but using cargoes of individual ships rather than averages reflects the variations in throughput and ties terminal handling costs more closely to activities of other decision makers at the terminal.

With these observations in mind, the total annual terminal freight handling cost incurred by the operator of the terminal may be expressed as:

$$C_{th} = FC_{th} + AVC_{th}[Q(a)] \tag{2}$$

where:
  FC_{th} = annualized fixed cost incurred in handling cargo by the terminal.
  AVC_{th} = average variable (or operating) cost per ton handled by the terminal.
  $Q(a)$ = annual tonnage handled at the terminal.
  a = mean arrival rate of ships at the terminal.

The mean arrival rate of ships is the average number of ships arriving in a period of time such as a day. Keep in mind that, as explained earlier, for simplicity's sake we have assumed all containerships entering the terminal are alike in their capacity and costs.

Storage costs. Like terminal freight handling costs, terminal storage cost may be divided into fixed and variable costs. As in the case of handling costs, one must assign an economic rent to the storage land and buildings. Care must be taken, however, because if we treat all storage costs as fixed, they then become "sunk costs," or overhead. This means they generally are not considered at all by the terminal operator in decision making or are considered a free good. In cost accounting terms, this means that the more one uses these facilities, the cheaper they become on a unit basis since the "overhead" is being spread over more units. This is valid accounting, but causes problems with cost modeling, since it does not reflect accurately the cost of processing and storing any given container. Decisions concerning an incremental container must be made using the incremental costs of handling that container.

Another problem in dealing with facilities that are owned by the terminal is that in making decisions with regard to changing terminal throughput, the terminal operator may choose to ignore storage capacity because it, too, is regarded as a "free" good. Hence, the possibility arises once again of exceeding terminal storage capacity, which may force the terminal operator to lease short-term capacity outside the terminal.[3]

With the above observations in mind, the total annual terminal storage cost incurred by the operator of the terminal may be expressed as:

$$C_{ts} = FC_{ts} + P_s \max[(QAST - SO), 0] \qquad (3)$$

where:
  FC_{ts} = annual fixed cost incurred in the storage of cargo by the terminal.
  AST = average stowage (cubic feet per ton) for storage of cargo.
  SO = storage capacity in cubic feet at the terminal.
  P_s = rental price per cubic foot of storage space incurred by the terminal operator.
  Q = annual throughput in tons of cargo at the terminal.
  $QAST$ = total annual cubic feet of cargo stored at the terminal.

The average stowage measurement (AST) is necessary to convert terminal throughput in tons to a measure of volume such as cubic feet. Such conversion is necessary to allow comparison of dissimilar cargoes or activities.

Ocean Carrier

Ocean carrier terminal costs are primarily a function of time—the time that a ship is in port. This "ship time" consists of queueing (or waiting) time and service time (time when the ship is actually being loaded and unloaded). A ship incurs queueing time when it enters the port harbor and must wait for a berth at the terminal.[4] Ship service time begins when a berth is available and the ship leaves the queue, and ends when the ship has had its cargo loaded and/or unloaded. Specifically, ship service time is berthing and unberthing time plus dock time or the amount of time the ship is tied up at the wharf.

Queueing time in a model such as this differs from the classical queueing model in that we include the time necessary for the ship to leave the queue and begin being serviced. In the classical queueing model, moving from the queue to the service point is considered to occur instantaneously. For break-bulk cargo, which may require several days or a week or more to process, that approximation may be close enough since an hour of berthing time is relatively insignificant. For containerships, however, which require only four to eight hours to be worked, an hour of berthing time may be significant. For all ships to be treated equally, the division between queueing and service time must occur when the ship leaves the channel and begins moving to a berth. To be more precise, queueing time is the waiting time from the moment that the ship anchors to wait for a berth until it begins to move from anchorage to a berth. If the ship does not have to wait for a berth, there is no queueing time.

While a ship is berthed, it incurs a variety of costs. Although a containership's crew does not normally participate in the loading and unloading process, they still must be paid. There are also the costs of depreciation of the ship itself and its capital equipment as well as maintenance costs and costs of supplies such as fuel. Since all of these costs are related to the time spent in port, the approach to building a cost model is to model ship time in port first and then convert that time to cost.[5]

The total queueing time for a ship at the terminal is a function of ship arrival rate, ship service time, and the number of berths at the terminal. The ship's total service time at the terminal is a function of the ship arrival rate and ship mean service time at the terminal. Hence, the annual cost (C_v) (the subscript "v" representing "vessel") incurred by ocean vessels (or ocean carriers or containership lines) at the terminal may be expressed as:

$$C_v = ATC_v[Z(a,N,s)] \qquad (4)$$

where:
ATC_v = average total (fixed plus variable) cost per unit of time for the ship at the terminal.
a = average arrival rate for ships at the terminal.
N = number of berths at the terminal.
s = average service time for ships at the terminal.
Z = total annual time (queueing plus service time) at the terminal for ships. The

expression Z(a,N,s) states that the total annual time depends upon the arrival rate, the number of berths and the service rate.

This cost equation has been simplified in much the same manner as our earlier equations. We have used average arrival rates and service times. The use of averages always eliminates ship-to-ship variations. If we were to use equation (4) in a simulation (or computer) model, we would have to be much more precise. Instead of average service time, we would need to determine a service-time probability distribution that reflects actual ship-to-ship variations. Such a distribution is typically a continuous probability distribution such as the negative exponential distribution shown in Figure 3.1. Simply using the mean may not generate the queues in the computer properly and could lead to poor decisions.

Similarly, ship arrivals may have to be described by an arrival distribution such as the Poisson distribution shown in Figure 3.2. Although such details are not necessary to understand the present discussion on cost modeling, one should be aware of them if a computer simulation project is undertaken.

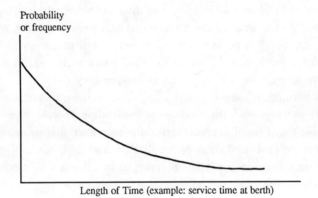

Figure 3.1. Example of a negative exponential distribution.

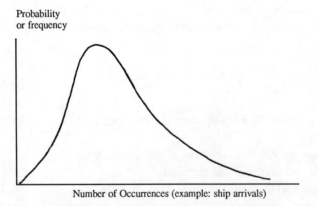

Figure 3.2. Example of a poisson distribution.

Inland Carrier

Inland carriers providing service to a marine terminal include long-distance trucks, railroads, and local drayage firms. Barges may be treated either as inland or ocean carriers depending upon how they are handled by the terminal. If they use the same wharves and equipment as ocean carriers (and, in essence, compete with the ocean carriers for their use), they should be treated as ocean carriers. If the terminal has special facilities such as canal access to the terminal, barges may be treated as inland carriers. For simplicity, we discuss here only truck traffic; inland barges and railroads would be treated essentially the same as trucks.

The terminal costs incurred by these carriers are a function of the time they take to process a load of cargo into the terminal. Basically, it is the elapsed time from when a vehicle enters the terminal until it leaves. A drayage firm, for example, usually measures its productivity by how many containers it can process per driver through the interchange process in a day. The longer each driver must remain on the terminal, the lower his productivity and the higher the cost per container.

In technical terms, the inland carrier interchange system in a large marine container terminal is a multichannel queueing system with well-defined channels. This means, for example, that as truck drivers enter the terminal with a container, they wait in a single line until an interchange gate is free. Shippers will often drop their cargo at a holding area away from the terminal and have local truck drivers haul the containers through the interchange onto the terminal, so the terminal costs are incurred by the drayage firm rather than the long-distance hauler. The terminal costs incurred by a truck drayage firm, for example, depend upon how many containers per day a driver can cycle through the interchange. That, in turn, depends upon the arrival rate of trucks, the number of gates, and the terminal service time incurred by trucks. The actual costs incurred by an inland carrier at the terminal include labor (hourly wages for the drivers) and equipment-related costs (mainly capital cost). The annual costs for such firms will be the average cost per hour of operating, times the hours spent hauling containers to and from the terminal. The annual costs (C_n) incurred by inland carriers at the terminal may be expressed in equation form as:

$$C_n = ATC_n[Z(a,G,s)] \qquad (5)$$

where:
- ATC_n = average total (fixed plus variable) cost per unit of time for the inland carrier at the terminal.
- $Z(\ )$ = total annual time (queueing plus service time) at the terminal for the inland carrier. $Z(a,G,s)$ means the total time on the terminal is determined by the arrival rate of trucks, the number of gates, and the service rate.
- a = mean arrival rate for the inland carrier at the terminal.
- G = number of interchange gates at the terminal.
- s = mean service time for the inland carrier at the terminal.

Again, the same approach may be used to model rail and inland barge costs.

Shipper

Shippers of cargo also incur costs at the terminal, namely inventory holding costs. These costs continue to accrue until the consignee actually receives the cargo. During the entire time the cargo is in the port or terminal area, the shipper has working capital tied up in the cargo and must pay a price for that. Thus the two problems that arise in determining terminal costs incurred by shippers are similar to those in the classical inventory management problem: (1) determining the average terminal inventory level, and (2) calculating the cost to the shippers of holding that inventory at the terminal.

The cost of holding the inventory at the terminal depends upon the weight of the cargo, its value, the holding cost per ton as a percentage of the value of the cargo, and the length of time it is held at the terminal.[6] Cargo may enter the terminal from two sources: (1) from ships (or barges) as they are unloaded and (2) from inland areas through the truck interchange or rail spur. Containers leave the terminal through the same gateways.

The first problem in modeling the shipper's cost is determining the value of the cargo. At a terminal where a small number of cargo types are dominant, it would be relatively easy to determine the average value of the inventory of the different types of cargo handled by the terminal. However, in a container terminal—where no one type of cargo is likely to dominate the inventory—it is more difficult to determine this average cargo value. For the larger terminal, it is probably sufficient to compute an average value per ton over the period of a year or so from terminal or government records. If the mix of cargo changes or is expected to change significantly, the value will, of course, have to be adjusted.

The holding cost would be expected to vary from shipper to shipper, but in general, the primary determinant of holding cost for inventory is the actual or implied interest on the cost of the inventory. The most straightforward approach for determining the real or implied interest rate is to determine a surcharge rate based upon local financial conditions and then add it to the current prime interest rate. The surcharge rate will depend upon acceptable internal rates of return in an industry or evaluation of risk in that industry. For example, if people in the industry agree that the surcharge rate is 4 percent of the value of the cargo and the prime rate is 12 percent, the holding cost would be 16 percent per year, or 16/365 percent per day. (Holding cost is based on the total number of days the cargo is in inventory, not the number of working days.)

The inventory on any given day at a terminal is the inventory at the beginning of the day plus the cargo both unloaded from ships (what we call "quayside activity") and brought in through the gate, minus the cargo both loaded on ships and taken out through the gate (what we call "gateside activity"). Modeling the change in inventory levels stemming from quayside activities is relatively easy, since it depends upon the arrival rate of ships and the amount of cargo loaded and unloaded.

Modeling the impact of gateside activities on inventory levels is a little more difficult since inland-carrier arrivals and departures are more independent of each other than quayside activities (since the same ship usually loads and unloads containers). Further, gateside activities are obviously related to ship arrivals and departures, but not as directly as quayside activities. Containers for a particular ship may arrive over a period of several weeks. In the case of a busy terminal, it would suffice to model the quayside and gateside activities independently. In a less busy terminal, it may be necessary to model gateside activities dependent upon ship arrivals.

The annual costs (C_f) incurred by shippers at the terminal may be expressed as:

$$C_f = (AHC)(AV)(AI)(365) \qquad (6)$$

where:
   AHC = average holding cost in terms of a percentage of cargo value at the terminal.
   AV = average cargo value per ton at the terminal.
   AI = average daily inventory in tons at the terminal.
   365 = number of days in a year.

Average inventory in tons (AI) is the arithmetic mean of the daily figures for ending inventory (calculated by adding beginning inventory plus cargo arrivals and deducting cargo departures). Quayside cargo arrivals and departures are a function of the ship arrival rate. Gateside cargo arrivals and departures are a function of their respective independent arrival and departure rates.

Harbor Support Auxiliary Firm

The costs incurred by harbor support auxiliary firms at the terminal consist of costs relating to towing, tugging, piloting, and berthing of ocean vessels. Modeling these costs can be difficult since they are incurred by a relatively large number of diverse firms. These costs include a combination of fixed and variable costs. The annualized fixed costs consist of the allocation of the capital investment in equipment such as tugboats. The allocation depends upon the life of the equipment and the accounting rules used to depreciate it. The actual allocation will change from year to year.

Annual variable costs (VC_b) incurred by harbor support auxiliary firms at the terminal depend upon the number of ships that are berthed at the terminal during the year. A simple formula for calculating VC_b is:

$$VC_b = (\text{operating cost/hour}) \times (\text{handling time/ship}) \qquad (7)$$
$$\times (\text{annual number of ships})$$

The operating cost per hour includes fuel and wages and is generally tabulated by the operators of the harbor service vessels (or their accountants). The handling time per ship and the number of ships are likely to vary widely. These may be estimated by making use of the average handling time per ship and the average number of ships handled by harbor support auxiliary firms at the terminal for a specified time period, such as a year. Alternatively, a service time function and a ship arrival/departure rate may be used. This latter approach allows for random fluctuations in ship arrivals and results in a more realistic, if more complex, model. The total annual costs (C_b) incurred by harbor support auxiliary firms at the terminal may thus be expressed as:

$$C_b = FC_b + AVC_b[H(s)][S(a)] \tag{8}$$

where:
- FC_b = annualized allocation of fixed cost of berthing assistance at the terminal.
- AVC_b = variable operating cost per hour (average variable cost) of berthing assistance at the terminal.
- $H(s)$ = ship handling time in hours as a function of the mean ship service rate (s) of berthing assistance at the terminal.
- $S(a)$ = annual number of ships receiving berthing assistance as a function of the mean ship arrival rate (a) at the terminal.

In more technical terms, the term H(s) could represent, for example, an exponential distribution with a mean ship service rate of "s". The mean ship service rate is the average number of ships that can be serviced at a terminal for a specified time period. The term S(a) could represent a Poisson distribution with a mean ship arrival rate of "a". Figures 3.1 and 3.2 illustrate these distributions.

Government

Terminal costs incurred by government will vary widely depending upon what services government offers. We assume in our example that they include expenses for dredging and maintaining harbor channels and for providing and maintaining harbor navigational aids. All of these costs are relatively fixed over a wide range of terminal activity. Channel dredging costs consist of both the cost of initially dredging a channel to a specific depth and the cost of maintenance dredging.

To determine an annual harbor cost for dredging, the initial costs of dredging must be allocated over time. In determining these allocations, one must take care to use a depreciation or allocation technique that reflects the useful life of the initial dredging. Further, if two or more ports or terminals share a harbor, the annual cost allocations of the initial dredging must further be allocated among the terminals that share the harbor.

Maintenance dredging is an operating cost, so it is charged in the year in which it takes place. Main channel dredging may have to be allocated among terminals along that channel, whereas the cost of dredging related to a specific terminal should be fully allocated to the terminal that benefits. Navigational aids may be treated in the same way as dredging. Initial navigational aid costs need to be allocated over years and among the terminals that benefit, whereas maintenance navigational aid costs are charged in the year in which they occur.

What distinguishes harbor costs from other costs previously discussed is that they are generally borne entirely and directly by a governmental agency (as contrasted to subsidies for ports or terminals) and are a "free good" to the terminals in the port. Harbor dredging and provision of harbor navigational aids, for example, may be performed by the government as part of the infrastructure for commerce, and the terminals in the harbor receive a "fringe benefit"; alternatively, they may be performed for a specific terminal. Our simplified model assumes only one terminal in the port. Based upon the above, the annual costs (C_g) incurred by government in the provision and maintenance of a harbor waterway for the terminal may be expressed as:

$$C_g = D + M \qquad (9)$$

where:
- C_g = annual port waterway costs incurred by governments at the terminal.
- D = annual maintenance dredging cost and annual allocation of initial dredging cost incurred by the government at the terminal.
- M = annual maintenance cost for navigational aids and annual allocation of initial navigational-aid cost incurred by government at the terminal.

COST SIMULATION

Simulation (or computer) models have a broad usefulness not normally found in pure analytical models.[7] They may be constructed for an existing marine container terminal or for a proposed (or hypothetical) marine container terminal. They allow the evaluation of a wide range of policies and decisions against an equally wide range of criteria. A properly constructed simulation model can answer most of the "what-if" questions a manager or analyst may ask. For example, a simulation model based on the cost model we have constructed could answer questions concerning the impact of a decision by one of the six terminal decision makers on its cost at the terminal as well as the impact of this decision on the costs of the other decision makers at the terminal. In Chapter 4, we discuss the uses of simulation models in more detail.

If we combine all the cost equations we have developed so far into one representing the total annual costs (CT) of activities at the terminal, the total costs may be expressed as:[8]

$$CT = C_{th} + C_{ts} + C_v + C_n + C_f + C_b + C_g \qquad (10)$$

In other words, CT (or equation 10) represents the sum of equations (2), (3), (4), (5), (6), (8), and (9). If the functional forms, probability distributions, and parameter values for the individual equations are known, then a computer (or simulation) model for CT (or equation 10) may be constructed.

Once the computer (or simulation) model has been constructed, it can be run to predict the costs to be incurred by one or more decision makers at the terminal from a change in the operating situation at the terminal. A shipper or containership line, for example, may decide to shift business from one terminal to another. The simulation model could predict the cost impact on the other participants at the terminal.

In addition to cost predictions, for a change in operations, the CT simulation model could also be used to predict operating statistics such as ship and gate arrivals; ship berthings, loadings, and unloadings; the number of ships in the queue; terminal inventory; and various average statistics.[9] In other words, it allows the various decision makers to predict the impact of a decision ahead of time. Alternatively, it allows them to see the impacts of different possible decisions when they are in the process of choosing a particular course of action.

CAPACITY MEASURES

Operating cost is not the only basis for decision models in container terminals. In addition to terminal cost simulations, a marine terminal operator may, for example, also use terminal capacity measures in his planning process. Capacity may be measured in many different ways, and we discuss the concept in several different contexts throughout this book. By whatever measures they use, many terminal operators believe they are over capacity or near their capacities to handle containers. In this chapter, we discuss capacity measures that may be used in constructing terminal models.

Specifically, such capacity measures may be used by the terminal operator for deciding whether or not to expand his terminal facilities. Because of the importance of capacity planning, we discuss it here in more detail. We give summaries of some types of capacity models with comments on each.

Definition

Although capacity is a concept that is often used in transportation planning, there is no agreement on what capacity means, and a number of different capacity measures appear in the transportation literature. Transportation capacity has been defined both in terms of an utilization rate as well as in terms of the size of a facility; it has been used as an engineering concept as well as an economic concept; and it has been used as a short-run as well as a long-run concept.

Capacity defined as a utilization rate has appeared both as an engineering concept and an economic concept in the transportation literature. Engineering capacity refers to the maximum capacity (or throughput) that physically can be handled by a given facility under certain conditions. As an engineering concept, capacity has been classified as: (1) design capacity, (2) preferred capacity, and (3) practical capacity.

Design capacity (or maximum capacity) is the maximum utilization rate for a given facility. For example, the design capacity for a storage area of a marine container terminal is the maximum number of containers that can physically be stored in the area.

Preferred capacity is the utilization rate for a given facility beyond which certain utilization characteristics or requirements cannot be obtained. For example, the preferred capacity for a given stretch of highway may be defined as the number of vehicles that use the highway per hour beyond which a given rate of speed for the vehicles cannot be maintained.

Practical capacity is the maximum utilization rate for a given facility under normal or realistic conditions. For example, the practical capacity for a shipside container crane at a marine container terminal may be measured as the maximum number of containers that the crane is expected to load and unload from a ship per hour under normal working conditions.[10]

Economic capacity may generally be defined as that capacity that can be physically handled by a given facility while satisfying an economic objective of the given facility. In satisfying the economic objective, this utilization rate may also be referred to as the "optimal utilization rate" for the facility. Economic cost capacity is the utilization rate for a given facility beyond which unit costs begin to rise, or the utilization rate at which unit costs are minimized. Thus the economic objective of the facility is to minimize unit costs. The economic cost capacity for a marine container terminal may be defined as that throughput (or number of containers passing through the terminal) per year at which the unit cost per container incurred by the terminal is at a minimum. In Figure 3.3, N* number of containers represents the economic cost capacity of the marine container terminal. Unlike engineering capacity, relatively little attention has been given to economic capacity in the marine terminal capacity literature.[11]

Short Run Versus Long Run

As an utilization rate for a given facility, capacity is a short-run concept, since the facility is assumed to be of a fixed size. Rather than measuring capacity as an utilization rate, however, it also has been measured in terms of the size of the facility. If such a measurement is used, the transportation planner is usually concerned with whether to change the facility's size; in other words, the planner is concerned with investment planning. If the size of the facility is allowed to change, capacity then becomes a long-run concept. This distinction is often unclear in the literature.[12]

An Engineering Capacity Model

Ideas about engineering capacity are a little less fragmented because of works such as Hockney's and Whiteneck's *Port Handbook for Estimating Marine Terminal Cargo Handling Capability*.[13] The handbook considers marine container terminals as well as various other types of cargo-handling marine terminals. Specifically, the handbook presents a modular method for estimating the capability of a given marine terminal where

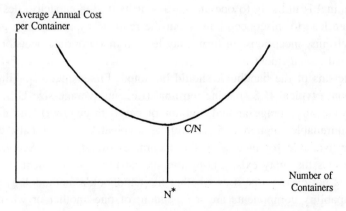

Figure 3.3. Economic cost capacity of a marine container terminal.

capability is defined as the number of tons of cargo transferred in one year given all the conditions under which the work must be done. The handbook presents a design of a marine terminal module that handles a typical quantity of a given class of cargo throughput. A terminal module is defined as a "typical" single-berth terminal designed to provide estimates of annual throughput values for specific classes of cargo. This marine terminal module may be employed by the user of the handbook to estimate the capability (or annual throughput) of a given marine terminal. For each terminal module appearing in the handbook, the following information is provided: a typical value for annual cargo throughput, a sketch depicting the principal physical features of the terminal being modeled, a list of throughput element values, a set of cargo throughput components, and a set of modifiers for the components.

In determining the capability estimate for a given marine terminal, the handbook estimates the capability of the various components of the marine terminal: ship/apron transfer capability, apron/storage transfer capability, yard storage capability, storage/inland transport transfer capability, and inland transport unit processing capability. The capability estimate for the marine terminal that is selected from among these five estimates is the estimate of lowest value. The capability component with the least value is the constraining capability component of the terminal (or "choke point") and hence an estimate of the maximum annual throughput (or capability) that the terminal can handle under normal working conditions. With this latter interpretation, it thus follows that the handbook provides estimates of the practical capacities of marine terminals.

The handbook presents a simplified approach for determining capability estimates of marine terminals that handle various types of cargo. Other methodologies for estimating engineering capacities of marine terminals have tended to use sophisticated analysis techniques that incorporate queueing theory, stochastic theories, and simulation techniques. In using the handbook to obtain capability estimates, an understanding of these sophisticated techniques is not necessary. Further, modifiers exist for adjusting terminal modules in order to increase the accuracy of their capability estimates.

Other methodologies for estimating engineering capacities of marine terminals generally seek to estimate the design capacity of these terminals, the maximum annual throughput when a terminal is operating always at a maximum rate of efficiency. Since a marine terminal is unlikely to operate always at its design capacity, design capacity estimates often lead to misconceptions about the realistic capability of a given marine terminal. Such misconceptions, in turn, may lead to inappropriate decision making by marine terminal management.

Three criticisms of the handbook should be noted. First, since capability estimates are based upon a typical U.S. marine terminal (i.e., an average-size U.S. marine terminal, providing an average annual amount of cargo throughput), the estimates are likely to be unreliable for non-U.S. as well as atypical U.S. terminals, even though modifiers are available for increasing the accuracies of estimates. Second, since the interdependencies that may exist among the six capability components are not stated explicitly, it is not clear to the user whether the handbook's methodology is assuming that these capability components are independent of one another or whether interdependencies are considered. More accurate capability forecasts could be obtained if these interdependencies were stated explicitly and thus known to the user.

Third, the handbook does not seek to provide an estimate of the optimal throughput of a marine terminal. In a competitive environment, a marine terminal is not only concerned with whether it can handle a given amount of cargo, but also whether it can compete with neighboring marine terminals for such cargo. It is one thing to know whether it can physically handle a given amount of cargo and another to know whether it can compete for such cargo. A marine terminal is more likely to be able to compete for cargo if it can lower the costs and hence the prices it charges for its services.

In addition to considering engineering capacity models such as the practical capacity model (where the maximum utilization rate is estimated for a marine terminal under normal conditions), marine terminals in a competitive environment should also consider economic capacity models. In utilizing the latter, the optimal throughput for a marine terminal can be estimated that corresponds to the economic objective or objectives of the marine terminal. For an example of an economic cost capacity model for a marine container terminal, see the appendix to this chapter.

SUMMARY

The activities at a marine container terminal are generally under the control of six distinct groups of decision makers. These decision makers include: (1) the terminal operator, (2) the ocean carrier (or containership line), (3) the inland carrier (such as rail and truck), (4) the shipper of cargo, (5) harbor support auxiliary firms (such as tug and pilotage firms), and (6) government. The activities of these decision makers at a marine container terminal incur costs; further, the activities (or decisions) by one decision maker at a terminal may cause a change in the costs incurred by other decision makers at the terminal.

The sum of the costs incurred by the various decision makers represents the total terminal costs incurred by terminal decision makers. From these and other data, we may construct a computer (or simulation) model representing the total terminal costs. With the computer model considering the terminal costs of individual terminal decision makers as well as cost linkages among these decision makers, the model can be run to investigate "what-if" questions concerning the impact of a decision by a terminal decision maker on its terminal costs as well as on the terminal costs of the other decision makers.

Capacity (or throughput) for a marine terminal may be measured: (1) as a utilization rate or in terms of the size of the terminal facility, (2) as an engineering concept or an economic concept, or (3) as a short-run or a long-run concept. Engineering capacity refers to the maximum throughput that a terminal can physically handle under certain conditions. Economic capacity refers to the throughput that physically can be handled by a given facility while satisfying an economic objective. Marine terminals have historically used engineering capacity measures. In a competitive environment, however, economic capacity measures should be considered, since a marine terminal in such an environment is not only concerned with whether it can handle a given amount of cargo but also whether it can compete with neighboring marine terminals for such cargo.

APPENDIX: AN ECONOMIC COST CAPACITY MODEL

In the following discussion, the total cost incurred by a marine container terminal (i.e., the terminal operator) is the sum of its handling cost, storage cost, administrative cost, and cost related to inland transport.[14] Specifically, the annual total short-run cost (C) incurred by a marine container terminal may be expressed as:

$$C = CH + CS + CA + CI \qquad (1)$$

where:
 CH = annual berth handling cost incurred by the given marine container terminal.
 CS = annual storage cost incurred by the given marine container terminal.
 CA = annual administrative cost incurred by the given marine container terminal.
 CI = annual cost related to inland transport incurred by the given marine container terminal (i.e., cost incurred by the terminal related to outbound and inbound movements of containers by such inland carriers as railroads and trucks).

The annual berth handling cost (CH) incurred by a given marine container terminal is the sum of capital and labor handling costs. Specifically, CH may be expressed as:

$$CH = CHB + CHE + CHL \qquad (2)$$

where:
 CHB = annual capital cost of berths incurred by a given marine container terminal.
 CHE = annual capital cost of berth equipment (ship-to-shore cranes, tractors, etc.) incurred by a given marine container terminal.[15]
 CHL = annual labor handling cost at berths incurred by a given marine container terminal.

The costs, CHB and CHE are fixed costs; CHL is a variable cost or

$$CHL = (WHL * \Sigma PT_i) \qquad (3)$$

where:
 WHL = labor cost per wharfside crane hour at the given marine container terminal.
 PT_i = the number of wharfside crane hours incurred by the ith containership at the given marine container terminal.
 ΣPT_i = annual number of wharfside crane hours actually incurred at the given marine container terminal.

The number of wharfside crane hours (PT_i) incurred for the ith containership may be expressed in terms of the following general function:

$$PT_i = f(Z_i, N_i, ND_i, PTR_i) \tag{4}$$

where:
  Z_i = size of the ith containership measured in deadweight tons (DWT).
  N_i = number of containers loaded and unloaded to and from the ith containership at the given marine container terminal.
  ND_i = distribution placement on the ith containership to which and from which containers (N_i) are to be loaded and unloaded.[16]
  PTR_i = number of wharfside cranes utilized in working the ith containership.

Further,

$$N_i = NK_i + NJ_i \tag{5}$$

where:
  NK_i = number of stacked containers loaded to and from the ith containership at the given marine container terminal.
  NJ_i = number of chassis containers loaded to and from the ith containership at the given marine container terminal.

The annual storage cost (CS) incurred by a given marine container terminal is the sum of chassis and stacking storage costs. Specifically, CS may be expressed as:

$$CS = CSK + CSJ \tag{6}$$

where:
  CSK = annual cost for storing containers by stacking incurred by a given marine container terminal.
  CSJ = annual cost of land area devoted to the storage of containers on chassis incurred by a given marine container terminal.

The annual stacking cost, in turn, may be expressed as:

$$CSK = CSKD + CSKE + CSKL \tag{7}$$

where:
  CSKD = annual cost of land area devoted to the storage of containers by stacking incurred by a given marine container terminal.
  CSKE = annual cost of stacking equipment (such as the transtainer crane) incurred by a given marine container terminal.[17]
  CSKL = annual labor stacking cost incurred by a given marine container terminal.

The costs, CSJ, CSKD, and CSKE, are fixed costs; CSKL is a variable cost. Annual labor stacking cost (CSKL) may further be expressed as:

$$CSKL = (WSL * \Sigma TTB_i) \tag{8}$$

where:
  WSL = labor cost per transtainer crane hour at the given marine container terminal.

TTB_i = number of transtainer crane hours incurred for the ith containership at the given marine container terminal.

ΣTTB_i = annual number of transtainer crane hours incurred for containerships at the given marine container terminal.

The number of transtainer crane hours (TTB_i) incurred for the ith containership may be expressed in terms of the following general function:

$$TTB_i = g(Z_i, NK_i, NDK_i, TTR_i, PT_i) \qquad (9)$$

where:

NDK_i = distribution placement in the stacking storage area of the given marine container terminal to which and from which stacked containers (NK_i) of the ith containership are stacked and unstacked.

TTR_i = number of transtainer cranes utilized in working the ith containership.

The annual administrative cost (CA) incurred by a given marine container terminal is a fixed cost and is the sum of administrative facility, equipment, and labor costs. Specifically, CA may be expressed as:

$$CA = CAF + CAE + CAL \qquad (10)$$

where:

CAF = annual cost of buildings and land facilities devoted to terminal administration by a given marine container terminal.

CAE = annual cost of equipment devoted to terminal administration by a given marine container terminal.

CAL = annual cost of labor devoted to terminal administration by a given marine container terminal.

The annual cost (CI) related to inland transport is the cost incurred by the given marine container terminal in providing inland transport transfer (in other words, providing for outbound movement of containers from the terminal by such inland transport carriers as rail and truck carriers) and in providing inland transport processing (in other words, providing for inbound movement of containers into the terminal by such inland transport carriers as rail and truck carriers).[18] This cost may be expressed as the sum of facility, equipment, and labor costs incurred by the terminal in providing the above inland transport service into and out of the terminal. Specifically, CI may be expressed as:

$$CI = CIF + CIE + CIL \qquad (11)$$

where:

CIF = annual cost of buildings and land facilities devoted to inland transport service incurred by a given marine container terminal.

CIE = annual cost of equipment devoted to inland transport service incurred by a given marine terminal.

CIL = annual cost of labor devoted to inland transport service incurred by a given marine container terminal.

The costs CIF and CIE are fixed costs; however, CIL will include both fixed and variable cost components. Specifically, CIL may be expressed as:

$$CIL = a + WIL*TTI \quad (12)$$

where:
- a = annual fixed labor cost (such as labor cost associated with interchange and inspection) for inland transport service incurred by a given marine container terminal.
- WIL = labor cost per transtainer crane hour for inland transport service incurred by a given marine container terminal.
- TTI = annual number of transtainer crane hours for inland transport service incurred by a given marine container terminal.

The annual variable cost of labor for inland transport service incurred by the terminal is represented by WIL*TTI. Therefore, it is assumed that the annual variable labor cost for the terminal is directly related to the annual number of transtainer crane hours incurred by the terminal in providing inland transport service. The annual number of transtainer crane hours, in turn, may be expressed in terms of the following general function:

$$TTI = h(NK) \quad (13)$$

where:
NK = annual number of stacked containers (i.e., number of "20-foot equivalent unit" containers) handled by a given marine container terminal.

The objective function for determining the economic cost capacity of the marine container terminal may be written as:

$$C/N = CH/N + CS/N + CA/N + CI/N \quad (14)$$

where:
$N = \Sigma N_i$ = annual number of "20-foot equivalent unit" (TEU) containers handled by the given marine container terminal.

Hence, in minimizing the annual unit cost of the marine coontainer terminal (C/N) with respect to "N", we obtain the annual economic cost capacity (N*) of the marine container terminal, the annual number of TEU containers handled by the terminal that will minimize the terminal's annual unit cost per TEU container handled. By substituting equations (2)–(3), (6)–(8) and (10)–(12) into equation (14), we obtain the following objective function for obtaining the economic cost capacity (N*) of a given marine container terminal:

$$\begin{aligned}C/N = {} & CHB/N + CHE/N + WHL * \Sigma PT_i/N + CSJ/N + CSKD/N \\ & + CSKE/N + WSL * \Sigma TTB_i/N + CAF/N + CAE/N \\ & + CAL/N + CIF/N + CIE/N + a/N + WIL * TTI/N \end{aligned} \quad (15)$$

Since N = NK + NJ (where, NJ is the annual number of chassis containers handled by a given marine container terminal and NK is the same as defined previously), it follows that by using NK and NJ as the choice variables for equation (15), not only can the economic cost capacity (N*) for a given marine container terminal be determined but also the optimal distribution of this economic cost capacity according to type of container, NK* and NJ*, can be determined. If the specific functional forms of the above equations are known, classical optimization theory (using calculus) may be used to obtain solutions for NK* and NJ*.

Notes to Chapter 3

1. The specific services provided by governments vary widely by country, and by states and cities within the countries, as does the accounting for the costs of these services. For example, in major harbors in the United States, the federal government maintains the channel depth as a free service to the port. In some ports in Great Britain, the costs of maintaining the channel must be recovered by the terminal operators in their pricing structures.

2. Applied Systems Institute and Maritime Administration, *Usage Pricing for Public Marine Terminal Facilities,* **Vols. I and II** (Washington, DC: U.S. Government Printing Office, 1981).

3. Jansson and Shneerson consider storage capacity to be a short-run constraining factor. However, we take the alternative view. Considering storage capacity to be a short-run constraining factor implies that ships could be refused access to a terminal because of storage limitations. However, it is more likely that a terminal would adjust to short-run capacity problems by leasing off-terminal storage space rather than risk antagonizing carriers or shippers. See J. O. Jansson and D. Shneerson, *Port Economics* (Cambridge: MIT Press, 1982).

4. For a discussion of queueing theory and queueing time, see D. R. Cox and W. L. Smith, *Queues* (London: Chapman and Hall, 1961).

5. A similar approach is found in Jansson and Schneerson, *Port Economics*. For further discussion of carrier costing, see W. K. Talley, *Transport Carrier Costing* (New York: Gordon and Breach Science, 1988).

6. One of the few studies to consider the cost to the shipper of his freight's moving through a port is a study by Kendall. In Kendall's model, however, this cost is measured only as that cost related to the time when the freight is in storage. In Kendall's model, the shipper's port cost (C_f) as related to ship size is expressed as follows:

$$C_f = (QSVI/200)/QY$$

where,
 QS = tonnage handled from a single ship.
 QY = total annual tonnage to be handled in the port.
 V = value of commodity per ton.
 I = percentage rate of return expected on the money capital invested in the freight during storage.

A shortcoming of Kendall's model is that the shipper's freight costs related to loading and discharging times (times other than when the freight is in storage) were not considered. For further discussion, see P. M. H. Kendall, "A Theory of Optimum Ship Size," *Journal of Transport Economics and Policy,* 6: 128–146.

7. In general, analytical models allow the calculation of a single, optimal solution to a problem. They require the rigorous mathematical description of all relevant functions in a port. A simulation model allows randomness, such as ship arrivals and service times, but provides the ability to test various scenarios. Simulation models also provide guidelines for decision making as opposed to the single optimal solutions of analytical models.

8. Examples of alternative marine container terminal models are found in: E. C. Frankel, "Algorithmic Approach to Container Terminal Design," *Annual Proceedings of the Transportation Research Forum,* 22: 180–187 (1981), and P. Schonfeld and S. Frank, "Optimizing the Use of a Containership Berths," *Transportation Research Record,* No. 984, pp. 56–62 (1986).

9. A discussion of marine container terminals and simulation models is found in E. G. Frankel, "Modeling Container Terminal Operations: An Application of Stochastic Probabilistic Simulation," *Proceedings of the World Conference on Transport Research,* 1: 864–875 (1986).

10. The U.S. Bureau of the Census uses both preferred and practical capacity in constructing utilization rates for a number of U.S. manufacturing industries.

11. An exception is W. K. Talley, "Optimum Throughput and Performance Evaluation of Marine Terminals." *Maritime Policy and Management,* 15: 327–331 (1988). For further discussion of economic capacity, see G. W. Wilson, *Economic Analysis of Intercity Freight Transportation* (Bloomington: Indiana University Press, 1980).

12. In order to distinguish between capacity as short-run and long-run concepts, Jansson and Shneerson refer to the former as fixed capacity and the latter as capacity investment. When the optimal size of a facility has been determined, Fernandez refers to this determination as the optimal investment in capacity. Increases in the size of transport facilities have been referred to by Borins as capacity expansions. Discussions of optimal investment in port capacity are found in a paper by Schonfeld and Frank and de Weille and Ray. In their study, the total system cost related to a single-berth marine container terminal is minimized with respect to the number of berth cranes to obtain the optimal number of such cranes. The total system cost is the sum of berth cost, cost of cranes, cost of storage yards, labor cost of crane gangs, cost of ships in port, and cost of containers and their cargo. The cost of ships in port is cost incurred by ship owners while the ship is in port and includes such costs as ship depreciation, crew, fuel, and insurance expenses. The cost of containers and their cargo is cost incurred by the owners of containers and cargo while they are in port. The cost of cargoes relate to the economic lifetimes of perishables and rapidly obsolescing items. In de Weille and Ray, the berth occupancy rate at a given port is simultaneously maximized with respect to the number of berths and the average service time per ship to obtain the optimal number of berths (or berth capacity) and the optimal average service time per ship. [S. F. Borins, "The Effect of Pricing Policy on the Optimal Timing of Investment in Transport Facilities," *Journal of Transport Economics and Policy,* 15 (1981): 121–133; "The Effects of Non-Optimal Pricing and Investment Policies for Transportation Facilities," *Transportation Research,* 163 (1982): 17–29; "The Economic Effects of Non-Optimal Pricing and Investment Policies for Substitutable Transport Facilities," *Canadian Journal of Economics,* 17 (1984): 80–98. J. E. Fernandez, "Optimum Dynamic Investment Policies for Capacity in Transport Facilities," *Journal of Transport Economics and Policy,* 17 (1983): 267–284. P. Schonfeld and S. Frank, "Optimizing the Use of a Containership Berth," *Transportation Research Record,* 984 (1986): 56–62. J. de Weille and A. Ray, "The Optimum Port Capacity," *Journal of Transport Economics and Policy,* 8 (1974): 244–259.

13. L. A. Hockney and L. L. Whiteneck, *Port Handbook for Estimating Marine Terminal Cargo Handling Capability* (Washington, DC: Office of Port and Intermodal Development, Maritime Administration, U.S. Department of Commerce, 1986).

14. The cost model for the terminal operator in this Appendix is consistent with the terminal-operator cost model in the body of this chapter. The former is a disaggregated form of the latter. Specifically, the annual cost (C_{ti}) associated with the operator of terminal facilities in the chapter

equals the annual total short-run cost (C) of a marine container terminal in the model in this appendix. The general structures of the models differ because of purpose of use. The former model is constructed for the purpose of determining the economic cost capacity of a marine container terminal. The latter model was constructed so that it could be adapted to a terminal cost simulation model.

15. Since such equipment as yard hustlers and yard chassis may also be used in the terminal's storage activity, the cost of the terminal's equipment must be allocated between the terminal's berth-handling and storage activities.

16. This variable measures the distances between the container locations on a containership to which and from which containers are to be loaded and unloaded. Greater the distances, greater will be the number of wharfside crane hours required to load and unload a given number of containers. A network formula may be utilized to determine ND_i.

17. In addition to transtainer cranes, top loaders and straddle carriers might also be used as stacking equipment.

18. Transfer refers to the transfer of containers in the possession of a marine container terminal to inland transport carriers. Processing refers to the activity of a marine container terminal in receiving containers being transported to it by inland transport carriers.

Chapter 4
EVALUATING TERMINAL OPERATIONS

In this chapter, we look at some problems frequently faced by terminal operators and some of the analytical and management techniques available for dealing with these problems. For purposes of organization, we group the problems into six categories: space utilization, capacity management, inventory management, line balancing and efficiency, yard management, and integration of operations. Although these categories are arbitrary, they provide a convenient framework for our analysis. Any categorization such as this is bound to contain overlaps, so some of the problems we discuss in one category could have been considered as part of several other categories instead. Our purpose, however, is not to be overly concerned about categories, but to describe the tools available to the terminal manager.

SPACE UTILIZATION

Space utilization has to do with how efficiently the physical land area on the terminal is used. The term "efficiently" is one that has common usage without a common understanding of what it means. Part of what we do in this chapter is present analytic techniques that allow the terminal management to define and measure efficiency.

Space utilization has traditionally referred to how much of a terminal's space is occupied by containers. Various statistical measures have been used to determine utilization, such as the maximum space occupied at any given point during a time period, or the mean space occupied during a time period. Space is utilized productively if it is occupied by containers or loaded chassis that are generating revenue as a result of their being on the terminal. Space is not utilized productively if it is vacant or is occupied by something not generating revenue—machinery, buildings, or even junk and trash. Note that although machinery and buildings may be necessary to operate a terminal, they are nonproductive in terms of revenue since their space could be used instead to store containers that generate revenue.

The efficient utilization of space becomes critical when the terminal operators perceive that they are approaching the capacity of the terminal in terms of space. At this point, pressures often intensify to increase the amount of land the terminal has or to somehow improve operations to increase the amount of space available at any given time. A clearer picture of capacity utilization, however, may be provided by a thorough analysis of what exactly is occupying the space on the terminal.

Space utilization is not a static problem. For example, the theoretical maximum container space on the terminal is the number of stack locations available times the height of the stack. Viewed in this way, most terminals are underutilized. Few terminals could even approach this theoretical maximum. Storing containers on chassis rather than in a stack would be one reason. The actual capacity is affected by a number of factors. Containers must be stacked methodically to allow the loading of vessels as efficiently as possible. In addition, containers do not simply sit on the terminal; they are constantly arriving and departing, both through the gate and by water. Finally, if productive use implies the generation of revenue, different types of containers may generate different amounts of revenue.

Dwell Time Analysis

The first step in the analysis of space utilization should be a "dwell time" analysis. It goes straight to the core of space utilization—how long containers remain on the terminal. Technically, dwell time is the time between a container's arrival on the terminal and its departure. Every hour or day a container remains on the terminal, it occupies a slot that could be used to service another container. The operational objective for the terminal should be to keep the dwell times as short as possible. Shorter dwell times mean that a greater throughput is possible for a terminal with a fixed amount of land. In other words, space is being utilized more efficiently.

Dwell time tends to be affected by a number of factors related to the characteristics of the containers. Part of the analysis for any given terminal should be aimed at determining these factors. The authors performed an analysis of approximately 38,000 containers over a three-month period at a terminal generally considered to be efficient. We discovered the important factors affecting dwell time to be (1) whether the containers were being exported or imported, (2) how soon import containers were permitted to leave the terminal by customs or other inspectors, (3) whether the containers were loaded or empty, and (4) whether or not the containers were "transient" (had no water leg to their movement).

Over the three-month period, import containers had a mean dwell time of about 8.5 days. In other words, the average import container remained on the terminal occupying a space for more than eight days, a surprisingly long time. Even more dramatic was the distribution of those dwell times. Over 1,000 of the containers remained on the terminal more than a month, and almost 300 remained on the terminal over 100 days. The underlying problem became obvious from an analysis of the dwell times of import containers after they received a pass to leave the terminal. The mean dwell time after the pass was issued was less than one day, and only four containers remained more than 30 days after the pass had been issued. It was taking an average of eight days to release containers so they could be removed from the terminal.

Further research would be needed to see what specifically was causing the delay. If it is a procedural matter, then it may be possible to make the procedures more efficient. If it is something beyond the control of the terminal operator, such as importers not paying their bills promptly, it may be necessary to increase the charges for containers remaining on the terminal for long periods of time. If the terminal allows free storage time after arrival, either because of tradition or contract, this may be another area for potential change. In other words, if it is difficult to reclaim the space, at least it should be made to generate more revenue.

This same terminal had slightly fewer export containers over the same three-month period, but the export containers had a mean dwell time of just under three days (as compared to 8.5 days for the imports). Part of this difference may be explained by the system itself. Most export containers are targeted toward a particular vessel with a scheduled sailing date. Shippers will normally schedule containers to arrive at the terminal as close as possible to the sailing date. Some terminals encourage this by guaranteeing fast processing of export containers. One terminal at the Waltershof Container Center in the Port of Hamburg guarantees 20-minute service from gate to vessel for precleared containers. In other words, a precleared container may arrive as late as 20 minutes before the vessel stops loading and still get on board. Import containers, on the other hand, arrive on a schedule, but usually do not have as rigid a schedule for departure.

Measuring "C-Days"

Two problem areas became apparent at our case study terminal. Empty containers made up about one-third of the total number of containers passing through the terminal, and transient containers were about one-sixth of the total. The empties had a mean dwell time of over 16 days, and the transients had a mean dwell time of nearly 12 days. In addition, the distributions of dwell times were dispersed or spread out. For example, over 1,500 of the empties had been on the terminal more than 30 days.

The measure the authors developed to determine the impact of a container on terminal space utilizatoin is "C-days." One C-day is one container on the terminal for one day. As noted earlier, the theoretical maximum is the number of container slots on the terminal, times the stack height, times the number of days in a time period. Again, this is far greater than the practical maximum. C-days, however, provide a much more accurate picture of the impact of various types of containers on terminal resources than the measures mentioned earlier. For example, in the study we discuss here, empty containers accounted for 27 percent of the total containers on the terminal during the three months, yet they accounted for 59 percent of the C-days. In other words, by usual measures, empties did not seem to be a problem. Using the C-days measure, however, they were shown to consume nearly 60 percent of the terminal's space. In another comparison, the total number of empties was about half the number of imports, but the 200 empties remaining on the terminal more than 100 days consumed more C-days than the 13,000 imports that had received passes to leave.

Dealing with empties and transient containers. If a terminal has sufficient space, storing empties may be a convenient method of generating revenue. For terminals with space at a premium, a high percentage of C-days consumed by empties can be a persistent problem. Empties have one characteristic, however, which makes a number of solutions possible—they weigh very little compared to a loaded container. This means that they can be stacked higher than loaded containers (for example, six high rather than four), and they may be handled with less expensive equipment such as side-loaders. They may also be stored on or off the terminal in areas that do not have the reinforced surface necessary to support the weight of loaded containers.

Transient containers (those with no water leg) also can provide revenue opportunities for a terminal with excess space. On our subject terminal, as noted earlier, about one-sixth of the containers were transient; those transients, however, consumed nearly 28 percent of the C-days. Since transients by definition have no water leg to their movements, it could be possible to keep them off the terminal altogether. (Customer relations could be a bit strained, however, if there are no other comparably priced accommodations for them locally.) The point is that problems with the availability and utilization of space may be caused by empty or transient containers and thus may be solvable without interrupting the flow of loaded containers.

Dealing with containers on chassis. A terminal with a pure chassis operation or one with a mixed chassis and stack configuration could perform a similar analysis. Determining the dwell times and C-days is independent of how the containers are stored on the terminal. Using the data on dwell times and C-days to measure space utilization is a bit different, however. Containers stored on chassis have several obvious, but important, differences from those stored in stacks. One is that since a container stored on a chassis is only "stacked" one high, there is no need to worry about the stacking sequence; the need will never arise to "dig out" one container from under another. Another difference is that containers stored on chassis do not require heavy equipment such as straddle carriers, side loaders, and transtainers to remove them from storage and transport them to shipside. This means that they do not have to be grouped according to ship or sailing date (although doing so might still improve the speed of loading) as they are likely to be in a stack operation. It does mean, however, that the information system containing the storage locations must be more accurate and faster than that necessary for a stack operation to allow the timely transfer of the container from storage to the ship.

What this means for land utilization is that a space is a space when it is reserved for a chassis, whereas in a stack, a space may be identified by import or export, shipper, vessel, or other distinction. Only one chassis-mounted container may be stored per space, and the space need not be identified otherwise (except, perhaps, for refrigerated containers and containers containing hazardous materials). The only significant variation is the length of the space; in other words, can it accommodate a 40- as well as a 20-foot chassis? The space utilization at any given time is the percentage of parking slots for chassis that are occupied. Empty chassis can become a problem if a terminal accumulates too many, but empty chassis can normally be stacked quite high in a minimum space (including unpaved areas or on racks), so they normally need not play a major part in the analysis.

CAPACITY MEASUREMENT

We discussed terminal capacity measurement in engineering and economic terms in the previous chapter. Here, we discuss capacity measurement as it is faced by terminal managers on a day-to-day basis. Capacity is not only the number of containers the terminal can hold in its present configuration, but also how many it could hold under varying configurations.

Capacity Trade-Offs

Actual capacity is a series of trade-offs with other aspects of terminal management. Placing containers in stacks, for example, is the most efficient way to occupy physical space on the terminal. It is the least efficient, however, in moving containers to and from ships. Storing containers on chassis, on the other hand, is very efficient for moving containers to and from ships but is land-intensive (it requires more land per container). It also may cause a bottleneck at the wharf since drivers must remain with the containers until they have been picked up by the loading crane. A similar bottleneck could develop if yard chassis are used to move containers from the stack to the wharf. Straddle carriers use land more intensely than chassis (they require less land per container), can stack to some degree, and may drop containers beside the ship in a sort of work-in-process inventory. Although they are cheaper than transtainers, more of them are required, and they are much more expensive than hustlers and yard chassis for transporting containers from the stack to the ship. At Hamburg's Waltershof Center, for example, one terminal is relatively land-rich and operates with straddle carriers. Its neighboring terminal, only a few hundred meters away, is relatively land-poor and operates with transtainers.

At the case study terminal mentioned earlier, the authors performed a complete inventory of stack and chassis storage locations. The terminal was land-rich relative to the number of containers handled. Our calculations indicated that each 40-foot chassis space was equivalent to 7.5 TEUs of stack space, if containers were stacked three high. This number is 7.5 rather than 6 because containers on chassis cannot be stored as close to one another as in a stack, and drivers need maneuvering room when moving chassis in and out of their spaces. Although we have not conducted a formal study of the trade-off between straddle carriers and the other forms of storage, it appears that each 40-foot chassis space is equivalent to about four TEUs of straddle carrier space, if the straddle carrier containers are stacked two high.

The amount of capacity required on the terminal is a function of the rate at which containers arrive at the terminal and the amount of time they remain (or dwell time). The management of dwell time through the measurement of C-days is the most effective method of controlling capacity on the terminal. The purpose of controlling capacity is to maximize the revenue generated by the space on the terminal, and C-days provide a direct connection between the actual use of the land as yard managers deal with it on a day-to-day basis and the revenue generated by that land.

Throughput Analysis

Capacity is not just the number of containers that may be stored on the terminal, however. It also refers to the number of containers that can be processed in a given period of time, such as a day. This may be looked in several different ways. One is throughput. The information typically available in a terminal MIS makes it practical to define throughput as occurring when a container leaves the terminal. (In industry, what we have called dwell time is sometimes referred to as throughput time.) Throughput may be measured on a container terminal by looking at how many containers depart the terminal each day. Although this is not a complete measure of handling capacity since containers need to be received on the terminal and moved when they are there, it does give us a picture of the variation in the day-to-day capacity required. At our case study terminal, we discovered the distribution shown in Table 4.1.

The mean throughput for the 64 sample days was about 613 containers per day. The standard deviation was 113 containers, indicating that throughput was concentrated in a fairly narrow range. The terminal managers could expect that on more than half the days, between 550 and 700 containers would leave the terminal. This means that, although the capacity requirements could fluctuate widely (from 350 to 900), most of the time they stayed in a fairly narrow range. This information could have implications for scheduling, hiring practices, and capital expenditures. A similar analysis at any given terminal could show the same results or radically different ones since the statistical results depend heavily on local conditions.

INVENTORY MANAGEMENT

Another way of looking at the containers on the terminal is as inventory. Standard inventory management techniques are not generally applicable since containers and chassis,

Table 4.1.
Container Throughput (containers per day)

Number of containers	Number of days
350–399	3
400–449	2
450–499	7
500–549	5
550–599	11
600–649	12
650–699	12
700–749	4
750–799	3
800–849	4
850–899	1

in manufacturing jargon, are serialized; they are identified and managed by serial number since each is unique and they generally are not interchangeable.

Many principles of inventory management are relevant to a container terminal, however. For example, a primary concern of the terminal manager should be the accuracy of the computer record of containers, since modern inventory management depends heavily on accurate computer records. Because containers and chassis are serialized, the location of each as well as the fact that both are present on the terminal are important and must be recorded accurately. In a mixed media terminal (one using both stack and chassis, for example), knowing the storage medium (stack or chassis) for any given container is also important. There may also be a difference between the day-to-day record of container locations needed for container management (the yard manager needs to know exactly where each container is located) and the historical record needed for analysis of terminal utilization and efficiency (which is less interested in exact locations and more concerned with dwell times and C-days).

Taking Inventory

A terminal that uses a manual container management system or that has a computer system that does not give immediate, accurate, up-to-date information about container locations may wish to perform a regular inventory of the containers and chassis on the terminal. Taking a complete physical inventory is difficult at most terminals because of the number of containers and chassis and because of the constant activity.

A more effective method for managing inventory is a technique known as "cycle counting." On the terminal, one would not be as concerned about the counting as about verifying whether a storage location contained the container or chassis identified in the computer record. What is stored in 1 to 5 percent of the locations on the terminal would be verified each week in accordance with a verification schedule that checked every location at least once per year. This would insure that over the cycle, every location would be accurate in terms of what is stored there. Cycle counting allows the yard managers to vary the frequency of counts according to importance of the containers. Locations for loaded containers probably should be counted or verified more often than locations for empties, for example. Cycle counting also spreads the workload over the entire year and does not require the terminal to be shut down for a physical inventory.

Because of the serialized nature of containers, unlike ordinary inventory items, they cannot become lost in the system for long unless the shipper or consignee somehow forgets about them. Not having a good storage tracking system, however, can cause the terminal to expend a large amount of resources finding temporarily "lost" containers. The authors have visited terminals that regularly lose 5 to 10 containers per day and others that lose only 5 to 10 per year. Much of the difference in the number of containers "lost" at these various terminals can be traced to the effectiveness of the inventory management techniques used at the terminals as well as the applicability and accuracy of the information flows.

Another way of looking at container management and space utilization is to use a measure called inventory turns. Since containers arrive and leave every day, one must calculate the number of inventory turns using averages. The more inventory turns a terminal has per year, the more containers it can handle in the same space (or the less space it will need to handle the same number of containers). In our case study terminal, the average throughput was 613 containers per day, and the average container inventory was 3,282. This means it took an average of 5.4 days (3,282/613) to turn over the inventory. The terminal had, therefore, about 40 (220 working days/5.4) inventory turns per year. In industry, a traditional manufacturing facility would consider this a good number. A retail establishment such as a food store would consider this low; so would a manufacturer using newer techniques such as JIT. We do not have sufficient information to make a precise judgment about inventory turns on a container terminal, but experience indicates that 100 turns per year (or an average of about 2.2 days to turn over the inventory) should not be an unreasonable objective.

Bar Coding

Another technique for managing and tracking inventories is bar coding, which has been used successfully for a number of years in relatively clean environments. In relatively hostile environments (such as bar coding of railroad cars), the early attempts were not particularly successful. Recent developments in the coding itself, in bar code readers, and in methods of printing codes have overcome most of the initial problems. The primary advantage of bar coding is that information has to be recorded only once manually, and thereafter re-recording is done by machine. In inventory management terms, if the information is recorded correctly the first time, the transactions accuracy for all subsequent transactions is 100 percent. If the transactions are recorded manually instead, each subsequent recording has a chance of being done incorrectly.

Some applications of bar coding await the adoption of industry standards—for example, the bar coding of the containers themselves. There are some applicatons that can be implemented on a single terminal, however. Paperwork can be bar coded with stickers containing information on container numbers, shippers, and destinations. Containers and chassis can be tagged with bar codes containing similar information. Locations can be bar coded. When a container is placed in a particular location, the container code and the location code could both be scanned to enter the information into the computer. In stack operations, location codes could be scanned automatically as machinery passed over them. This could lead not only to more accurate data entry, but also to semiautomated operations. For example, instead of the normal procedure of a checker recording the location either manually or through a computer terminal, the location information could be scanned and transmitted to the central computer automatically. Alternatively, the checker could use a bar coding wand to enter the data without risk of human error instead of entering them through a key board. Details on bar codes and their application can be found in a rich literature of both books and journals.[1]

The use of microwave readers is an alternative to bar coding. The readers are small devices (15 cm square or smaller) that emit a code electronically. They are read when a special device is passed over them. This system has the same advantages as a bar coding system in terms of entering data into the MIS. It has the advantage of being less affected by a "dirty" environment since the reading of the data is electronic rather than optical. There are several drawbacks that may inhibit their use on a container terminal, however. Since, unlike bar codes, they are separate devices, they can become lost or damaged. They also require relatively expensive machines for programming and reading the devices. And, unlike bar codes, if the electronic reader malfunctions, there is no way to read the information manually.

LINE BALANCING AND EFFICIENCY

Line balancing and efficiency refer to two different phenomena. One is the absolute efficiency of any given operation on the terminal in isolation—for example, loading containers on a ship or removing containers from a stack. The other is the efficiency of an operation relative to other operations with which it is connected. Keeping capacities approximately equal at each component of the process is known as line balancing. Typically, a process, such as accepting a container on to the terminal, involves a series of operations.

Interrelationships among Terminal Processes

At one of the Waltershof terminals in Hamburg, for example, a container arriving at the terminal goes through three basic operations. The first is the interchange where the paperwork is completed and the container is accepted. The second is the yard control center where the container is assigned to a storage position and a straddle carrier to take it there. The third is the placing of the container in storage by the straddle carrier. The overall process can only operate as quickly or as efficiently as its least efficient or slowest component.

Both absolute and relative efficiencies are important to the terminal manager. A process may be in balance, but at a too low a level of efficiency. Increasing that process efficiency requires measuring and improving the efficiency of the individual components. Improving all components may not be necessary, however. Improving the least efficient component will increse the efficiency of the overall process.

Time-and-Motion Studies of Cycle Times

The most appropriate technique for dealing with line balancing and efficiency on container terminals is a form of time-and-motion study. This type of study involves gathering data on the cycle times of various operations. Cycle times are those times necessary to complete one individual operation. For example, if one is concerned about the efficiency of unloading a ship, data need to be gathered on the cycle time of the ship-to-shore crane, the cycle time of the hustlers or straddle carriers that transport containers between the crane and the storage areas, and, if appropriate, the cycle time of the equipment (such as transtainers or side loaders) placing the containers in storage.

The cycle time of the ship-to-shore crane is the time required to remove one container from the ship and place it on a chassis or on the ground for a straddle carrier. The cycle time for the trailer truck or the straddle carrier is the time between the moment it picks up one container and the moment it picks up the next. Thus a cycle always starts and ends at the same point in the process.

Although the average cycle time is important, just as important is the variation in the cycle times. A process that is perfectly balanced in terms of average efficiencies will still experience inefficiencies caused by variations in the cycle times. The wider the variations, the greater the potential for these inefficiencies.

The technique for gathering data is relatively straight forward. What is required are a watch, a spreadsheet for recording the data, and, perhaps, binoculars to remain unobtrusive and avoid biasing the data by one's presence. The primary data one should collect are the exact hour, minute, and second of important events for each piece of equipment. For example, if a ship-to-shore crane is unloading containers from the ship and placing them on yard chassis, which then take them to a transtainer to be stacked, the data for the yard chassis should be the time each receives a container and the time each returns to the ship ready to receive another container. The differences between the two times will tell the analyst both the times the hustler driver must wait at the ship to receive a container and the time required to unload the container at the stack and return to the ship. A second analyst should also be stationed at the stack to record the arrival times of the hustler driver and the departure times after the stacking equipment has removed the containers. Similarly, recording the times the ship-to-shore crane places each container on a chassis will tell the analyst the cycle times of the crane.

Data on each ship-to-shore crane should be recorded separately as should the data on each truck and straddle carrier. The cycle time should be recorded for every move with notations when unusual events occur, for example, if a ship-to-shore crane must wait because no truck or straddle carrier is available to remove the container it has unloaded. One should also record all observed causes of nonproductive activity. For a ship-to-shore crane unloading a ship, for example, the only productive moves are those that move containers from the ship to the ground transportation—anything else is nonproductive. Nonproductive activities may include lateral moves by the crane, changing the spreader bar, waiting for trucks or straddle carriers, carrying men or equipment on and off the ship, rest breaks, changing operators, and breakdowns. Some nonproductive activity may be necessary, but one must know what the machine does at all times in order to manage line balancing and efficiency effectively.

Similar data need to be collected for all the other operations in the process. The amount of data that need to be collected is substantial. The actual results will depend upon a number of factors such as the characteristics of the ship, the crew unloading the ship, whether the ship is being primarily loaded or unloaded, the terminal, the time of day, the time during the shift, the weather, and the season. The data should be collected over a complete operation, such as the unloading and reloading of a selected ship. The entire process of loading and unloading the vessel and the associated yard operations need to be examined simultaneously. This is important to determine interrelationships such as the causes of delays. For example, if a wharfside crane must wait to unload a container, the person recording data in the stack may find that the transtainer was involved in nonproductive activity at the same time. Failing to study the whole process at once would cause the analyst to miss these relationships.

An operation that, by its relative inefficiency, slows down the entire operation is known as a bottleneck. This relative inefficiency may result from several sources. This component of the process may not have the capacity to keep up with the other operations. It may, for instance, have slower machines, or too few machines, or too little labor to operate at the pace of the others. On the other hand, it may simply have too much nonproductive time relative to the other operations. Sometimes a bottleneck is temporary. This results from variations in the cycle times. An operation becomes a temporary bottleneck if it experiences a slow variation in its cycle time, when the other operations in its process are experiencing fast variations in their cycle times. If there is significant variation in each of the component operations, the bottleneck will appear to move from one place to another as the cycle times vary.

These types of studies will produce a wide variety of results because of the variations in ships, crews, and the other factors we mentioned earlier. Terminal managers face an even wider variety of actions they may take to respond to problems identified through such studies. For example, an inefficient operation in a process may be balanced by increasing its capacity. Capacity may be expanded by increasing the capital equipment available, the labor available, or the utilization of that capital or labor. On the other hand, shortening the cycle time may increase the capacity without increasing resources. On a terminal, this may result from decreasing nonproductive activities, or, in some cases, changing layouts, shortening distances, or changing traffic patterns. For instance, if straddle carriers must travel long distances to store the containers removed from the ship, shortening the distance by changing the traffic patterns or terminal layout will reduce the cycle time and improve the efficiency. Time-and-motion studies can be complex to analyze but have the advantage that the close observation of the system required to collect the data will often suggest obvious ways of improving the efficiency of the operation.

In the previous section we deal with the interactions of the various components of the operations on the terminal. What we have not discussed is how decisions by the ship lines may impact on terminal operations. In the next chapter, we discuss alternatives in ship design. Here we point out a few examples of how ship design and terminal operations interact. An obvious example is that a ship with no superstructures, masts, or cranes between hatches to interfere with the operation of the ship-to-shore crane can be unloaded more quickly than a ship that requires more nonproductive moves by the wharfside crane. Large container ships may offer both advantages and disadvantages to the terminal operator. If the ship is large enough (and has modern ballasting capabilities), weight balancing becomes a less significant factor in the stowage plan and the yard manager does not have to be as precise when prepositioning containers in the stack or staging area. The containers may be picked from the stack in the most efficient sequence—that which minimizes nonproductive moves by the transtainer. The same large ship can cause problems for the yard manager, however, because of the volume of containers that is loaded or discharge in a relatively short time. Managing the flow of containers through the terminal is more difficult when the containers arrive in large batches than when the flow is relatively smooth over time.

YARD MANAGEMENT

Central to the task of yard management at a marine container terminal is the flow of information into and within the terminal. In Chapter 2, we describe the nature of the management information system (MIS) that is necessary to accomplish the objective of more efficient container management. In the following discussion, we look at methods of evaluating the information system and at alternative forms of MIS.

Improving Terminal Interchange Information

The interchange at the terminal is the place where data on a particular container are either collected initially or are verified, if the terminal has received advance information. Accuracy is critical at this point, since a container could be temporarily lost if it is not correctly identified at the interchange function. Should inaccurate data become a problem, it may be attacked with the state of the art technological innovations we discussed earlier such as optical scanners for container numbers, bar coding of containers, bar coding of documents by shippers, or microwave encoders and readers. If necessary, the system may be constructed so that the interchange writer never directly enters or retrieves data in the system, preserving existing computer-related work jurisdictions.[2]

Second, unless the terminal is an all-stack operation (which few are), the interchange writer must either gather relevant information on all chassis or be able to distinguish between those that will leave the terminal and those that will remain. Even the most advanced system that identifies containers before they arrive on the terminal may not capture the same information on the chassis. This is because the chassis is merely a transportation medium, whereas the container is the item being transported. The system must be designed to compensate for the fact that the time required increases and the possibility of error is greater since the interchange writer must deal with two items (the container and chassis) instead of just the container.

Improving Information Flows

As noted in Chapter 2, no matter how sophisticated a terminal's computerized system, the results obtained from it cannot be any more accurate than the data that are entered into it. To ensure that a person enters data correctly, he or she must either have a vested interest in the data being correct or be conscientious, properly trained, and aware of the consequences of incorrect entry, or both. An indirect vested interest may be created in a competitive atmosphere such as that found in the mid-Atlantic U.S. ports or the northern European ports. Workers there can be reminded that errors cause dissatisfied shippers who may then switch their business to another port or terminal. When shippers change terminals, people lose their jobs. A more direct vested interest may be created by the threat of suspension or dismissal for employees who repeatedly generate too many errors. A properly configured computer system could help in identifying employees who are error prone.

Proper training of terminal employees is at least as important as discipline, because when throughput is increasing, employees with less seniority and experience often must be used (for example, as checkers who record the identity and location of containers). In short, just when the workload pressure and need for immediate accuracy increase, the terminal often must rely more heavily on less experienced, undertrained employees.

Disciplining an inadequately trained employee for doing a job improperly is generally ineffective and unfair. A training course for checkers or others involved in gathering data is an important component in improving the accuracy of data essential to more efficient yard management. Management trains people in technical skills, yet often assumes anyone can learn to record and enter data accurately. The training should include emphasis on the role of the employee in the overall operations of the terminal and the importance of absolute accuracy. Each trainee should undergo proficiency testing as a precondition for working in a job requiring data collection. As one might expect, testing as a precondition for employment is likely to increase interest in and attention to the training. Since yard configuration and operation change from time to time, some retraining for all employees is likely to become necessary. In addition, management should be alert for the possibility of using technological innovations such as those discussed previously to improve data accuracy.

Entering Location Data Faster

At terminals where unionized employees are not allowed to deal directly with the computer system, where the necessary technology is not available, or where the technology is available but simply has not been implemented, identifying inaccurate data in a timely manner is difficult because a "runner" must be used to transmit the data from the checkers to the computer entry point. Since it may take a runner an hour to make a circuit in a medium-size terminal, at least an hour passes before it is even possible to know if a checker has entered an invalid container number or location, or if the checker has assigned a container to a location already taken. In the terms defined in Chapter 2, the system is not real time, and both accuracy and applicability may suffer. If the information were known within minutes instead of hours, it would be possible to take immediate action.

In such situations, a nonunion supervisor may be placed in the yard in a vehicle with a portable mobile computer terminal. The supervisor could make a continuous circuit of the checkers and enter the information immediately, instead of sitting in container control waiting for a runner to deliver it. Any discrepancies could be flagged and given to the checkers to correct. Many errors would be corrected on the spot, and others within a relatively short time. This would enhance both the applicability and accuracy of the data, since it would be entered in the computer sooner, and errors would be corrected more quickly. The use of a container control supervisor in a vehicle with a terminal is a relatively unsophisticated solution, but it avoids dramatic changes in work rules and requires a much less sophisticated computer and data entry system than a system that inputs data directly into the computer. In other words, the system may not be real time, but the lag between data collection and availability has been greatly reduced, and it provides an alternative for terminals that do not have access to the latest sophisticated technology.

Radio Transmission of Data

The technology for linking a vehicle to the computer is relatively straightforward. A number of hand-held or mobile computer terminals are on the market. A radio link probably would involve placing an antenna at a central point in the yard and connecting the antenna to the container control facility housing the interface to the computer. There are terminals using such systems in Antwerp, Hamburg, New Orleans, and Norfolk.

There are at least three considerations in selecting a particular radio transmission system. One is the availability of appropriate frequencies. Marine terminals tend to be in metropolitan areas, so frequencies may be difficult to obtain. The second is the potential for radio interference *by* government or commercial transmissions. And third is the potential for interference *with* government or commercial operations. At one time, the ECT container terminal in Rotterdam perceived the frequency problem to be so serious that they decided to use an infrared communications system. Collection points were installed around the yard and were triggered when vehicles passed over them.[3] The system avoided the radio frequency problem, but the timeliness of the data suffered since it was not real-time collection.

A radio link to the computer also gives supervisory personnel real-time access to the computer data base. Without a computer radio link, they must use a system such as two-way radio to ask container control for information and wait until someone has accessed the data base and replied. If supervisors have hand-held or mobile terminals, they can make their own inquiries directly to the computer; in other words, they can be on-line. Consequently, they can spotcheck containers and chassis that appear to look "out of place" and make immediate corrections. Having a data base that is more accurate and timely also makes supervisory personnel more willing to rely on the computer outputs such as lists of vacant parking slots, rather than use manual systems as those mentioned in Chapter 2.

One way to increase the accuracy of location data (which results in faster retrieval and placing of containers) is to use systematic and logical location designations. For example, each location in a transtainer stack may be marked with a four-digit number. The first digit could represent an area of the yard, the second digit the row, and the third digit the space across the row. The fourth digit would indicate the vertical position in the space. In this way, anyone familiar with the system would know immediately the location of a container by its location number. The physical locations also must be clearly marked and lighted so they may be readily identified, even by the inexperienced. In some cases, color coding may help both to identify locations and designate proper traffic patterns. The same principles are valid for chassis and straddle carrier systems, although the details may vary.

In addition to the above location data, additional information can be entered into the MIS to make the stack management system even more effective. For example, within stack areas assigned to particular lines, containers (especially those for export) should be stacked according to characteristics such as voyage, size, port of destination, and sometimes weight. If this type of information were accurate and accessible from the data base (real time and on line), a supervisor could, for example, ask for a list of the locations of all containers destined for a particular port. Those containers that were out of position could then be moved when a transtainer was idle.

The ultimate manifestation of such a system is prestowing. Containers could be put in the stack so that they could be removed in the exact order needed to load a ship. This will require more container moves by yard personnel, but should minimize the moves necessary when the time comes to actually load the ship.

If the proper technology were available and had been installed, a supervisor would not have to remember all the stack information, but could retrieve it with a portable terminal. Since the designated areas would not be permanently fixed (unless the terminal served a small number of lines and had sufficient land capacity), a supervisor also would be able to modify the boundaries of these areas through a portable terminal as the need arose. A hard copy report would still be possible with a computerized system, but the data base would be easier to update and accessible to all supervisory and managerial personnel. Such a system lays the groundwork for more efficient stack management such as "immediate" or "nearly immediate" retrieval and delivery of containers from the storage area by the over-the-road carriers without the necessity of using a separate holding area.

Obtaining Accurate Information from Rail and Shipping Lines

In Chapter 2, we noted the potential data problems in dealing with railroads and ship lines—primarily the difficulty in getting accurate, applicable information. Different terminals will have different degrees of difficulty (if any at all) with rail and ship lines depending upon their locations and the specific railroads and ship lines they deal with. In the following discussion, we assume a "worst case" scenario to allow a full development of the topic, realizing that some terminals may have solved some or all of the problems noted.

Several approaches to dealing with the problems are available, some short-run and others requiring long-term cooperation of all the parties involved. An example of an immediate short-run solution in dealing with the problem of railroads failing to provide complete information on containers due to arrive at the terminal is to place a clerk at a nearby switching yard (if one exists) or even farther up the line to call ahead with the information. The cost of placing a clerk at the switching yard may be more than offset in savings of yard labor as a result of the improved information.

A typical and understandable solution to the problem of inaccurate information from the containership lines is for terminal operators and port authorities to maintain pressure on containership lines to provide timely information. As we pointed out earlier, however, only those convinced of their vested interest in doing so will provide accurate and timely information. For example, because containership lines have an obvious vested interest in good information, many of them have been convinced to provide advance manifest data. On the other hand, once the railroad has loaded a train with export containers, it has little to gain by transmitting timely information to the terminal. An example of an interactive vested interest in good information is a ship line that has a cooperative agreement with a railroad to operate a unit train. Since the containership line has a vested interest in accurate and applicable information, the containers arriving by rail should be well documented.

The ultimate soluton for capturing and maintaining accurate information is to design a system that provides the information without requiring additional action by those who have no vested interest in its accuracy or timeliness. A direct communications link among computers would be such a system, for example, because at the point where, say, the rail line's interest in the data ends, the data are immediately available to the marine terminal. In effect, all the participants would have access to the others' data bases. Such systems are technically feasible and are widely used within organizations. Their widespread adoption for data exchange among different organizations is primarily inhibited by concerns about computer and data security.

Analysis of Container Moves

Most of the labor and equipment on a marine terminal is used for moving containers. This means that a large proportion of the variable costs that the terminal management must recover through its rate structure stems from moving containers. Any changes in policies or operating procedures designed to make the terminal operate more efficiently need to be evaluated in terms of their impact on the number of moves they require. For example, if a terminal is all-chassis and the loading and unloading of ships is done by independent stevedores, the terminal itself needs very little labor and equipment. A stack operation, on the other hand, requires a heavy investment in equipment and personnel. The trade-off may be worthwhile in light of other factors such as the amount of land available, but an analysis must be performed to assess the exact ramifications of policy changes.

For one terminal studied by the authors, we summarized the accumulated "moves" of the containers in our sample. We defined a "move" as any movement of a container performed by terminal personnel, which, therefore, increased the cost to the terminal. Thus we excluded the parking of containers on chassis by over-the-road drivers and moves performed by independent stevedores.

At this particular terminal, 66 percent of the containers in the sample accumulated no moves, and 32 percent accumulated one move. Only 2 percent had more than one move, with the maximum being six moves. A container with no moves would have remained on a chassis throughout its stay on the terminal. Therefore, it appeared that two-thirds of the containers at this terminal remained on chassis. The proportion may actually have been higher, since the containers with one or more moves were not always moved immediately but spent part of their time on the terminal mounted on chassis.

The mean number of moves for all containers was 0.372. In other words, on the average, only one out of every three containers was moved. Suppose the terminal management wished to move more containers into the stack in order to improve space utilization (containers in the stack require less room than those mounted on chassis). This would require a minimum of one move by the terminal (into the stack); moving the container out of the stack to the ship would be the responsibility of the stevedore. The total number of moves would increase by about three times. Instead of one out of every three containers being moved, every container would be moved at least once. Another policy designed to speed terminal operatons might be immediate stacking supplemented by a buffer-staging area. In other words, containers would be moved from the stack to a staging area on or near the wharf to speed the loading of the ship. This policy would require at least two moves for each container, and would increase the total number of moves by a factor of six compared to the original situation.[4]

Policies such as the above two suggestions may result in large gains in space utilization (and thus capacity), but they must be balanced against the increased equipment and labor necessary to make the additional moves. The labor and equipment required are not likely to increase in the same proportion as the number of moves, however. The above policy requiring two moves, for example, is not likely to require six times the capital and labor of the original policy because of increased operating efficiency (which would be the reason for implementing such a policy). Even if operating efficiency were to double, however, the capital and labor requirement would have to triple in order to perform six times the number of moves. If terminal managers double or triple the amount of capital and labor required, they must also consider how they will recover the associated costs. If the intent is to increase the space utilization in order to increase the amount of business at the terminal, the required capital and labor may increase even faster than double or triple. If the increased cost is to be recovered through the rate structure, shippers and ship lines must feel they are gaining something in exchange for the increased cost. On the other hand, the increased capital and labor may be a substitute for additional land. Whatever the situaton, performing a moves analysis gives the terminal management the data to do a thorough financial analysis of proposed policy changes.

INTEGRATION OF OPERATIONS

One danger of performing narrowly focused procedure or policy change analyses is what may be called "analytic myopia" or "suboptimizing." Changes that improve part of the operation, in fact, may cause other parts of the terminal to operate less efficiently. One way to minimize this problem is computer simulation. Ideally, a computer simulation should model all the parts of the terminal operation, including their interactions. This would allow management to test the full effects of any proposed changes in terminal operations before investing time and money in making the changes. On the other hand, computer simulations are not trivial to construct, nor are they inexpensive. Once they have been constructed, however, they can allow rapid, inexpensive analysis of proposed decisions or actions. In order to provide a feeling for what a simulation can offer management, we describe two such models below.

Panagakos Model

This simulation model, the work of George Panagakos (most recently of the World Bank) while he was a student at MIT, was developed using the SLAM II simulation language. This language allowed the simulation model of a terminal to be developed using network techniques. The terminal is represented as a network through which containers flow—sometimes waiting (as in storage) and sometimes being serviced (as in being loaded aboard ship). Since Panagakos formulated the model when the PC-XT was the standard microcomputer, he simplified many of its aspects to allow it to run in a reasonable time (each run took about one hour). Recent improvements in computer hardware technology would allow the construction of a more complex model. In addition, improvements in simulation software such as SLAM II allow simulation models to be constructed much more quickly and efficiently.

One of the simplifications that Panagakos made was that he dealt only with an all-chassis terminal. This means the simulation did not consider the interactions between the chassis and the stack. The model was concerned primarily with the operation of the terminal and not with outside factors. Use of the model allows the analysis of increases in business at the terminal, different ship calling schedules, and the land, capital, labor mix necessary to operate the terminal.

Maritime Trade and Transport Program Model

A more complex model was formulated as part of the Maritime Trade and Transport Program at Old Dominion University. It included the interactions among all six participants at the terminal (as discussed in Chapter 3)—government, auxiliary services (such as tugboats), the ship lines, the shippers, the terminal managers, and the over-the-road haulers. (For more detail on the model, see Chapter 3.) It was a cost model designed to evaluate the total impact of operational policy decisions. Although it is primarily a model for evaluating operations, it had to be formulated in terms of costs in order to provide a basis for comparison among all the participants. A decision that simply shifted costs from one participant to another could be distinguished from one that lowered the overall costs of operating the terminal. The addition of an additional gantry crane, for example, might increase the costs of operating the terminal, but also might decrease the operating costs for the ship line and increase the utilization of the marginal wharf space. Policy changes at the interchange gate might impact that operation and also over-the-road carriers and work scheduling within the terminal itself.

An extension of this work was a simulation model formulated by one of the authors for the Virginia Center for World Trade. The model was constructed to study the impact of container traffic on noncommercial traffic patterns around the Hampton Roads metropolitan area and especially in the vicinity of the terminal. In addition, the model could test the effect of changes in noncommercial traffic patterns on the ability of containers to move in and out of the terminal. Constructing the model required extensive data collection on traffic patterns throughout the area, and especially on the routes used to move containers within the area.

Simulation Requirements

As stated earlier, constructing a simulation model is not easy or inexpensive. In order to make the model useful, the analyst must make simplifying assumptions about the interrelationships on the terminal. Judgments must be made about which relationships are important and which are unlikely to affect use of the model. A great deal of data must be available to allow modeling the relationships. Earlier we pointed out some areas in which statistical analysis may be beneficial; simulation modeling, however, requires considerably more detail than the techniques discussed earlier. Constructing the model itself is quite painstaking and requires considerable time for testing and validation. When the model has been completed, scenarios for testing must be carefully formulated. The time required to test the scenarios often may extend to weeks, so the model tends to be more useful for longer term strategic planning rather than for day-to-day decisions.

As an example of the strategic uses of a simulation model, the authors organized and participated in a conference, part of which was devoted to designing simulation scenarios for a medium-size terminal. The first scenario included a number of operational changes, such as stacking immediately from over-the-road vehicles, more frequent ship and barge arrivals, more frequent train arrivals, and performing simultaneously certain tasks that previously had been done in sequence. The second scenario was to evaluate the effects of different treatment of containers. For example, the operating rules for handling containers could be based on either the ship line transporting the containers, or on the characteristics of the container itself (such as import, export, or empty). The third scenario involved the evaluation of the impact of new equipment that would have different operating characteristics and speeds from the present equipment. The final scenario was "empty chassis off the terminal" that would experiment with alternative approaches to offsite chassis storage that varied cycle times, entry and departure procedures, or that treated chassis for different lines or different uses differently.

Simulation computer programs are available through commercial companies. For example, Terminal Consulting Holland (TCH) markets a container terminal simulation (CTS) model that runs on a microcomputer. Some of the problems they have identified as being within the scope of their simulation are: (1) stacking requirements, (2) the effect of berth length on the offered service, (3) the relation between the number of cranes, crane production, and service, and (4) consequences of alternative berth and/or crane allocation decisions. Again, these are strategic uses of simulation aimed at long-term capital investment decisions.

Although it is commonplace for a business wishing to construct a simulation to hire the expertise from a consulting firm or university, some in-house expertise must exist or be developed to allow the full utilization of the model. This may mean hiring someone with the expertise or retraining existing personnel. Ideally, the person should have a thorough knowledge of the terminal operations, the modeling process, and computers.[5]

SUMMARY

This chapter presents techniques for evaluating terminal operations. Except for simulation that requires some specialized knowledge, the techniques presented are easy to apply and interpret and are based on data likely to be readily available at most container terminals. We have considered the problems faced by terminal managers on a day-to-day basis—the utilization of the existing space and the control of the containers on the terminal.

Space utilization needs to be considered not just in terms of how much space is occupied, but also what occupies it and for how long. Dwell time and C-days analysis give the manager a clear picture of what is utilizing the space. In addition, looking at the containers on the terminal as inventory gives terminal managers a basis for using some of the recent techniques developed in industry for managing that inventory, such as cycle counting and bar coding.

Using established techniques of time-and-motion study in combination with more recent ideas on bottleneck analysis gives the manager a better overall picture of the terminal operation. Efficiency and line balancing allow a more economical focus of the resources available on the terminal. Finally, making sure that accurate and applicable information is in the hands of yard managers helps the terminal run more smoothly and efficiently.

Notes to Chapter 4

1. For example, the American Production and Inventory Control Society (Falls Church, Virginia) publishes a number of books on bar coding. *ID Systems* is a journal by Helmers Publications, Inc., which is dedicated to bar coding technology and applicatons. There is an annual ID Expo Trade Show and Conference in the United States. Many manufacturers and consultants offer bar coding equipment and training.

2. In some ports, management has chosen to restrict the direct access of unionized workers to the computers. This derives partially from the desire to avoid redefining job descriptions and work rules. Whatever the constraints under which the terminal is operating, however, there is technology available to improve data collection and maintenance.

3. A straddle carrier, for example, would have on-board equipment that would record movements by the carrier—containers moved, locations of those containers, and so on. Rather than transmitting these data directly to container control, they would be stored in the on-board equipment until the carrier passed over a collection point. A collection point would be buried in the yard and consist of equipment such as an infrared receiver and a transmitter wired directly to container control. This avoids radio transmissions, but the data are not transmitted to container control until the carrier happens to pass over a collection point. Depending upon the number of collection points and the path of the carrier, the data may be current or old by the time they are transmitted to container control.

4. If the original number of moves per container was 1/3 (or 0.333) and the new number is two per container, the increase is 2/0.333 or 6.

5. Those wishing a more complete discussion of the benefits of simulation modeling should consult one of the many books in the field. One of the best is *SLAM II Network Models for Decision Support* by Pritsker, Sigal and Hammesfahr (Englewood Cliffs, NJ: Prentice-Hall, 1989).

Chapter 5
SHIP TECHNOLOGY AND COSTING CONTAINERSHIP SERVICE

The preceding chapters focus on the marine container terminal. In this chapter and in Chapter 6, we turn our attention to the containership. This chapter presents a discussion of the containership itself as well as the costing of containership service. Chapter 6 examines the issue of containership routing and the problem facing the carrier of choosing between different all-water patterns as well as various intermodal networks among ports.

We begin by considering aspects of containership technology, primarily design and capacity. Then we discuss containership costs—vessel costs while at sea and in port as well as port charges. Finally, we examine optimal containership size by analyzing the impacts of route distance, terminal efficiency, and load centering on optimal size.

CONTAINERSHIP TECHNOLOGY

The historical development of container shipping and the cellular container system is well documented (see Chapter 1). In this section, we discuss containership design and measures of containership capacity.

Containership Design

The first generation of containerships were existing ships that were converted for the transport of containers. They were usually "self-sustaining" or "geared" in that they had their own lifting gear and thus could use any marine terminal berth that was available. For the past three decades, however, ships have been built specifically for the transport of containers. Many are nonself-sustaining (gearless) in that they do not have container lifting gear aboard. Some are completely cellular, meaning that all cargo space below decks has been divided into cells and thus is suitable only for containerized cargo. Some are equipped with cell guides that make the placement of containers on the vessel easier for the crane operator, thereby enhancing the handling speed of containers to and from the ship.

The carrying capacity of a vessel is enhanced when cargo is stored in containers because deck space may be utilized more extensively. Modern containerships often carry containers up to five high on deck, held in position by a combination of corner locking devices and lashings. By utilizing the deck, approximately 30 percent more cargo can be carried by the average containership than if the ship's holds alone were used.

Containerships may be classified as cellular, semicontainer, and RO-RO. The cellular containership (see Fig. 5.1) is the most efficient type for moving large volumes of containers. Unlike conventional vessels, the cellular containership normally does not carry lifting equipment. Thus it is dependent upon the marine terminal to have cranes for the loading and discharging of a ship's containers. The rationale for not carrying lifting equipment is that it (1) lowers construction costs, (2) minimizes the amount of cargo space sacrificed to make room for equipment, and (3) permits the vessel to be worked more rapidly, since wharf cranes (which are typically much faster than ship cranes) can move from hatch to hatch without having to be raised to get over the ship's cranes. It is assumed, of course, that such vessels will be used only on routes between ports that have appropriate lifting equipment.

Semicontainer or combination ships (see Fig. 5.2) are designed to carry other types of cargo (bulk or break-bulk), in addition to or instead of containers. Many semicontainer ships were designed under the assumption that the evolution toward containeri-

Figure 5.1. Cellular containership. (*Courtesy*: Evergreen Marine Corporation)

Ship Technology and Costing Containership Service   81

Figure 5.2. Semi-container (combination) ship. (*Courtesy*: Lykes Brothers Steamship Co.)

'zation would be gradual.[1] RO-RO vessels (see Fig. 5.3), as noted in Chapter 1, are designed to permit containers (as well as other cargo) to be driven on and off the ship. Since this inevitably involves sacrificing some space for internal ramps and decks, the RO-RO vessel is less efficient than the cellular containership in handling large volumes of containers.

There is considerable variation in the design of RO-RO vessels. In addition to roll-

Figure 5.3. Ro-Ro containership. (*Courtesy*: Virginia Port Authority)

on/roll-off capability, many RO-ROs also have lift-on/lift-off (LO-LO) capability, either by using their own cranes or those of the terminal. These vessels are sometimes referred to as LO-ROs. The proportion of cargo space available to LO-LO handling can vary. In some cases, the ship is equipped with cellular holds. In others, LO-LO cargo (non-containerized as well as containerized) can only be accommodated on deck. Furthermore, the shipborne ramps of RO-ROs can vary. Some have stern ramps, others have quarter or side ramps, and stern ramps may be straight, angled, or slewed.[2]

Containership Capacities

A containership (or any type of ship) may be described in terms of three types of ship capacity measures: (1) holding capacity, (2) handling capacity, and (3) hauling capacity.[3] Ship design as well as the characteristics of the port of call and the amount and composition of cargo will affect the values of these capacity measures.

A containership's holding capacity is the maximum number of containers (TEUs) that the containership can hold. For cellular containerships, evidence suggests that an increase in a ship's deadweight tonnage (DWT) will result in the same percentage increase in its holding capacity. (A ship's DWT is the number of tons of 2,240 pounds that it can hold.) In economics, the ratio of the percent change in one variable to the percent change in another variable is referred to as the measure of elasticity between the two variables. In a study by one of the authors, the elasticity between the holding capacity and the DWT of cellular containerships was found to be approximately one, thus confirming the perception that a given percentage increase in the size (DWT) of a cellular containership would be expected to result in the same percentage increase in the maximum number of TEUs that it can hold.[4]

A containership's handling capacity is the number of containers (TEUs) that can be loaded into or discharged from the ship per hour while in port. In the same study noted above, the elasticity between the handling capacity and the size (DWT) of cellular containerships was investigated and found to be 0.247. This elasticity indicates that a given percentage increase in the size (DWT) of a cellular containership would be expected to result in a smaller percentage increase in the number of TEUs that can be loaded into or discharged from the ship per hour while in port. Based upon the elasticity of 0.247, one would predict that a 10 percent increase (for instance) in the DWT of a cellular containership would be expected to result in a 2.47 percent increase in the ship's handling capacity. It is interesting to note that the elasticity of containership handling capacity with respect to ship size (DWT) of 0.247 is in conformity with similar containership elasticities of 0.19, 0.242, and 0.423 found by Jansson and Shneerson.[5] Further, it is in conformity with Thorburn's principle of 0.3 for the elasticity of ship handling capacity with respect to ship size for tramp ships.[6] These studies assume, of course, that the number and speed of the lifting equipment employed remain unchanged.

A containership's hauling capacity is measured in terms of container-miles per day. A containership's hauling capacity may be found by multiplying its holding capacity by its hauling speed (number of nautical miles per day).[7] Further, the elasticity of containership hauling capacity with respect to ship size (DWT) can be found by summing two elasticities: (1) the elasticity of containership holding capacity with respect to ship size (DWT) and (2) the elasticity of containership hauling speed with respect to ship size (DWT).[8] As stated previously, the former elasticity is approximately one for cellular containerships. In the above noted study, the estimate of the latter elasticity for cellular containerships was 0.091.[9] Since the sum of these two elasticities (or the elasticity of containership hauling capacity with respect to ship size) is greater than one, one would expect that a given percentage increase in a cellular containership's DWT will result in more than a proportional increase in the ship's hauling capacity.

CONTAINERSHIP SERVICE COSTS

We now turn our attention to the discussion of vessel costs incurred by containerships at sea and in port as well as port charges incurred in port.

Table 5.1 presents vessel operating cost (excluding fuel expenses), construction cost, and various characteristics of four different size cellular containerships. These data were obtained from the Maritime Administration (MarAd) of the U.S. Department of Transportation. In providing operating expense data on ships, it is normal practice for MarAd not to include fuel costs. The reason given is that fuel costs vary greatly according to a ship's itinerary; this variance also applies to ships of the same type. (In a study by one of the authors, fuel costs were found to be 49.6 percent of a containership's operating costs).[10]

Several conclusions can be drawn from the data in Table 5.1. First, when ship size increases from 600 TEUs to 2,450 TEUs (or 308 percent), annual operating cost increases by only 27 percent, suggesting that there are economies of ship size with respect to annual operating cost. Second, when ship size increases 308 percent, construction cost increases by only 161 percent, indicating that there are economies of ship size with respect to construction cost. Third, when ship size increases by 308 percent, the crew size remains relatively stable, indicating that there are economies of ship size with respect to crew size and, therefore, crew wages. Thus one may conclude that for the range of containership sizes considered in Table 5.1, cost economies of ship size exist for cellular containerships. However, if ship costs are related to the number of TEUs loaded on and discharged from containerships in port, a different conclusion may follow. In the next two sections, we relate the vessel costs of cellular containerships incurred at sea and in port to the amount of service (TEUs moved per day) provided by these ships.

Vessel Cost at Sea

In Figure 5.4, we depict the relationship between the daily cost per TEU incurred by a cellular containership at sea, referred to as the unit sea vessel cost, and the number of TEUs transported per day. A cellular containership's daily total vessel cost (operating and capital) at sea per TEU is plotted against the number of TEUs transported per day. As one would expect, the unit sea vessel cost curve EK in Figure 5.4 is negatively

Table 5.1.
Cost and Vessel Characteristics of Cellular Containerships (1984)

	Ship Size (TEUs)			
	600 TEUs	1,200 TEUs	1,600 TEUs	2,450 TEUs
1. Annual Operating Cost, US$ (Excluding Fuel Cost)				
Wage & allowances	1,195,000	1,195,000	1,195,000	1,330,000
Subsistence	71,000	71,000	71,000	78,000
Stores, supplies, & equipment	68,000	91,500	115,500	115,500
Maintenance & repairs	367,500	430,500	449,500	525,000
Insurance	240,000	299,000	344,000	413,500
Other expenses	31,500	31,500	31,500	42,000
Total	1,973,000	2,118,500	2,251,500	2,504,000
2. Construction Cost (Japan built in 1983, US$)				
	19,500,000	28,400,000	36,900,000	50,800,000
3. Vessel Characteristics				
Deadweight tonnage	12,000	20,000	28,000	42,000
Propulsion	Diesel	Diesel	Diesel	Diesel
Gross tonnage	13,000	23,000	41,000	40,500
L.O.A.	530	670	950	860
Draft	27	33	35	30
Horsepower	15,000	28,500	35,000	43,000
Speed (kts)	18	22	24	22
Fuel consumption (bbls/day)				
at sea	400	760	900	1,050
in port	50	70	90	125
Crew size (Liberian flag, Greek crew)	29	29	29	32

Assumptions: Operating days per year equals 350; 60 percent of operating days at sea and 40 percent in port.
Source: Maritime Administration, U.S. Department of Transportation.

sloped, thus indicating that there are cost economies of ship utilization at sea; unit vessel cost declines as the ship's load factor or percent of cargo space utilized at sea increases.[11] Since vessel capital costs are fixed, and many operating costs (wages, allowances, and insurance) do not vary with the ship's load factor at sea, one would expect a ship's unit sea vessel cost to decline as the ship's load factor at sea increases.

Further support for our conclusion of cost economies of ship utilization at sea for cellular containerships is found in a study in which the measure of service provided was ton-miles per day, rather than TEUs transported per day.[12] In this study, the elasticity between containership operating cost (including fuel cost) and ton-miles per day of service at sea was found to be 0.17, thus indicating that a 10 percent increase, for example, in ton-miles per day of service will result in only a 1.7 percent increase in containership operating cost.

In the above discussion, we analyze the unit sea vessel cost for a given cellular containership under the assumption that its load factor was changing. We now consider

Ship Technology and Costing Containership Service

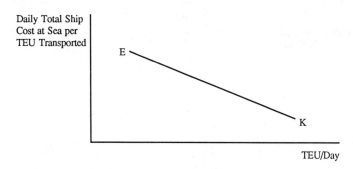

Figure 5.4. Cost economies of containership utilization at sea.

the unit sea vessel cost of cellular containerships under the assumption that the load factors (or percent of cargo space utilized) of the ships are the same. Hence, different size ships are required to transport the various numbers of TEUs at sea. In Figure 5.5, the daily total containership cost at sea per TEU transported is plotted against the number of TEUs transported per day for different size containerships having the same load factor. The unit vessel cost curve in Figure 5.5 is negatively sloped, thus indicating that there are cost economies with respect to increases in the number of TEUs transported per day at sea and ship size.[13] The symbol "S" in Figure 5.5 represents the size (DWT) of cellular containerships, where S3 > S2 > S1. Support for this conclusion is found in the elasticity between the hauling capacity and size (DWT) of cellular containerships. This elasticity (being greater than one) indicates that an increase in ship size will result in a more than proportional increase in hauling capacity. Thus with vessel costs not increasing in proportion to ship size (DWT) and with service provided at sea (or ship hauling capacity) increasing more than in proportion to ship size, unit vessel cost with respect to service provided at sea is expected to decrease with increases in service provided and ship size.

Vessel Cost in Port

In Figure 5.6, we depict the relationship between the unit port vessel cost (daily cost/TEU) incurred by a cellular containership in port and the number of TEUs moved

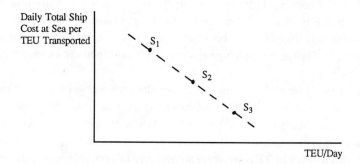

Figure 5.5. Cost economies of containership size at sea.

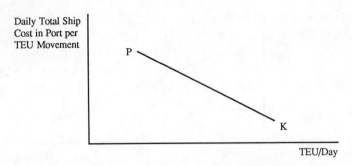

Figure 5.6. Cost economies of containership utilization in port.

(loaded on and discharged from the ship) per day in port. A cellular containership's daily total (operating and capital) vessel cost in port per TEU movement is plotted against its number of TEU movements per day. Data found in Gilman reveal that this plotted relationship (curve PK in Fig. 5.6) is negatively sloped, thus indicating that there are cost economies of ship utilization in port; unit port vessel cost declines as the number of TEUs loaded on and discharged from a ship in port increases.[14] The rationale for this conclusion is similar to the rationale given above for cost economies of ship utilization at sea for cellular containerships. Since much of the vessel cost incurred in port does not vary with the number of TEUs the ship loads or discharges, unit port vessel cost should be expected to decline as the number of TEU movements increases.

In our previous discussion, we concluded that cellular containerships exhibit cost economies at sea with respect to increases in the number of TEUs transported per day at sea and ship size. In port, there is evidence (although not conclusive) suggesting that cellular containerships exhibit cost diseconomies. For cellular containerships in the size range of 600 to 3,000 TEUs, Gilman found that, if ships experienced the same number of TEU movements per day, unit vessel cost (in TEUs/day) increased with ship size.[15]

If both the number of TEU movements per day in port and ship size are varying, however, it is possible for cellular containerships to exhibit either cost diseconomies or economies in port. If the elasticity between vessel costs incurred in port and ship size (percent change in port vessel cost/percent change in ship size) is greater than the elasticity between ship handling capacity per day and ship size (percent change in TEU movements per day in port/percent change in ship size), cellular containerships will exhibit cost diseconomies in port with respect to increases in their number of TEU movements per day in port and ship size. If the former elasticity is divided by the latter elasticity, we obtain the elasticity between vessel costs incurred in port and ship handling capacity per day (percent change in port vessel cost/percent change in TEU movements per day in port).

The elasticity will be greater than one if the elasticity between vessel costs incurred in port and ship size is greater than the elasticity between ship handling capacity per day and ship size. With the elasticity being greater than one, port vessel costs will increase more in proportion to the increase in ship handling capacity. Thus unit port vessel cost with respect to TEU movements per day in port will increase with increases in TEU movements per day in port and ship size (cost diseconomies in port).

If the elasticity is less than one, unit port vessel cost will decrease with TEU movements per day in port and ship size (cost economies in port). Based on the above discussed elasticity ratio, it follows that if cellular containerships exhibit cost diseconomies in port with respect to TEU movements per day, these cost diseconomies can be lessened or dissipated by efficiency improvements in the handling of containers in port.

Port Charges

In addition to vessel costs incurred in port, a containership also incurs port charges for services rendered by port authorities, stevedore companies, tug operators, and pilots. These port charges may include charges for pilotage, tuggage, dockage, wharfage, linehandling, stevedoring, vessel overtime, rental of terminal cranes, and number of containers loaded on and discharged from ship. Pilotage is a service rendered by a licensed marine pilot in guiding a vessel in or out of a port or harbor (or through other waters). Tuggage is a service rendered by a tugboat firm in the towing, tugging, or berthing of ocean vessels. Dockage is, in effect, the rental of space alongside a pier or wharf where the ship ties up. Wharfage is, in effect, rental for use of space on the wharf itself at which a ship is berthed. Linehandling is the tying and untying of a ship to or from a wharf. As described in Chapter 2, stevedoring is the service of loading or discharging a vessel's cargo. Vessel overtime is time over and above that for which the vessel was scheduled to be in the berth.

For the Port of Hampton Roads, for example, pilotage fees are based upon the overall length, breadth, and depth of the containership. Tugs charge on the basis of the net tonnage of the ship, and dockage and wharfage charges are based upon the gross registered tonnage of the ship and the net tons of cargo loaded and unloaded, respectively. The linehandling charge is a flat fee per vessel, and stevedoring costs are the product of the stevedore charge per gang hour and the number of gang hours incurred to work the ship. The vessel overtime charge is computed on an hourly basis. Rent for terminal cranes is the product of the rental charge per hour and the number of hours the equipment is used, and fees for container movements are the product of the receiving charge per container and the number of containers loaded and discharged.

Based upon a sample of 36 different-size cellular containerships that called upon the Port of Hampton Roads in 1985, the elasticity between daily port charges incurred by these ships and their size (DWT) was 0.279, after adjusting for other factors affecting the level of port charges.[16] This elasticity suggests that a given increase in ship size (DWT) can be expected to result in a less than proportional increase in daily port charges. Alternatively, unit port charges (port charges per DWT) can be expected to decrease as ship size increases, thus indicating port charge economies of scale for cellular containerships.

If unit port charges are based upon TEU movements incurred by cellular containerships per day, however, an opposite conclusion may follow. If the above elasticity between daily port charges and ship size (percent change in daily port charges/percent change in ship size) is greater than the elasticity between ship handling capacity per day and ship size (percent change in TEU movements per day/percent change in ship size), cellular containerships will exhibit port charge diseconomies with respect to their number of TEU movements per day and ship size. If the former elasticity is divided by the latter elasticity, we obtain the elasticity between daily port charges and ship handling capacity per day (percent change in daily port charges/percent change in TEU movements per day).

If the above elasticity ratio is greater than one, daily port charges incurred by cellular containerships will increase more in proportion to the increase in ship handling capacity; thus unit port charges with respect to TEU movements per day in port will increase with increases in TEU movements per day in port and ship size (port charge diseconomies). If so, these diseconomies can be lessened or dissipated by port efficiency improvements in the handling of containers.

If the elasticity is less than one, unit port charges will decrease with increases in TEU movements per day in port and ship size (port charge economies). If so, containership lines can take advantage of these economies by load-centering—utilizing larger containerships and calling at fewer ports (load-center ports) where cargo is congregated.[17]

OPTIMAL CONTAINERSHIP SIZE

In the previous sections, we discuss the responsiveness of cellular containerships' costs at sea, costs in port, and port charges with respect to ship size and TEU movements per day. These elasticities as well as the responsiveness of cellular containership size to marine terminal performance, load-center ports, and containership route distance have been considered by one study that investigates optimal containership size.[18] Specifically, cellular containership annual cost (at sea and in port) per number of TEU movements per day were minimized with respect to containership size (DWT) in order to determine optimal containership size. Both operating costs (including fuel) and capital costs were considered as well as port charges.

For a round trip route of 9,002 nautical miles and 13 round trips per year, the study estimated that the optimal containership size for this route was 36,500 DWT. For a round-the-world route of 25,374 nautical miles and assuming 4.7 round trips per year, the study estimated that the optimal size was 84,250 DWT. The rationale for why optimal containership size should increase with route distance is that as the distance of a round trip route increases, a containership will spend more days at sea and therefore can take greater advantage of the cost economies at sea.

The optimal containership size for the shorter route (36,500 DWT) is similar to the size of cellular containerships currently used on such routes. The optimal containership size (of 84,250 DWT) for the longer route is larger than the jumbo vessels (of size 63,402 DWT) used by United States Lines or Evergreen Lines on round-the-world routes in the mid-1980s. However, as Chapter 7 suggests, future jumbo's will be even larger, approaching and perhaps surpassing 80,000 DWT.

In addition to estimation of optimal sizes of cellular containerships, the study also reached conclusions in regard to containership cost economies and diseconomies. First, since vessel cost per TEU movement per day at sea (or unit vessel cost at sea) decreases as ship size (DWT) increases, cellular containerships exhibit cost economies of ship size at sea. Second, since vessel cost per TEU movement per day in port (or unit vessel cost in port) increases as ship size (DWT) increases, cellular containerships exhibit cost diseconomies of ship size in port. Third, since port charges per TEU movement per day increase as ship size (DWT) increases (based upon data utilized in the study), cellular containerships exhibit port charge diseconomies of ship size.

In addition to the investigation of ship cost economies and diseconomies, the study also investigated how responsive optimal ship size is to changes in containership handling capacity (the number of TEUs that can be loaded on or discharged from a containership per hour). Suppose the efficiency of marine terminals in loading and discharging containerships improves. If so, the handling capacity of containerships will improve, thus resulting in an increase in the elasticity of containership handling capacity with respect to ship size. Table 5.2 presents estimates from the study of optimal containership sizes for the 9,002 nautical-mile route if the elasticity of containership handling capacity is varied with respect to ship size. As is evident, optimal containership sizes are quite sensitive to relatively small changes in the elasticity of containership handling capacity. For example, the 20 percent increase in the elasticity from 0.250 to 0.300 results in a 52 percent increase in optimal containership size.

The study also found that load centering results in increases in optimal containership size. Load centering occurs when a significant concentration of containers exists at a smaller number of ports. With the greater concentration of containers at ports, a larger number of containers on average will be loaded on or discharged from containerships per port call, or the average number of ship-to-shore crane hours per port call will increase. The study found that increases in these crane hours per port call results in an increase in optimal containership size.[19]

SUMMARY

Containerships may be classified as cellular, semicontainer (combination), or RO-RO vessels. The cellular containership is the most efficient type of containership for moving large volumes of containers. Three capacity measures have been used to analyze the capacity of containerships: (1) holding capacity, (2) handling capacity, and (3) hauling capacity.

Table 5.2.
Elasticity of Containership Handling Capacity and Optimal Containership Size

Elasticity Coefficient	Optimal Containership Size (DWT)
.150	19,700
.200	29,500
.250	36,500
.300	55,300

Cellular containerships exhibit cost economies of ship utilization (or load factor) at sea and sea cost economies with respect to increases in the number of TEUs transported per day at sea and ship size. Also, cellular containerships exhibit cost economies of ship utilization in port; unit port vessel cost (cost/TEU) declines as the number of TEUs loaded on and unloaded from a ship in port increases.

Although not conclusive, some evidence suggests that cellular containerships exhibit cost diseconomies with respect to increases in their number of TEU movements per day in port and ship size—unit port vessel cost (cost/TEU) increases with increases in TEU movements per day and ship size. Evidence from one port suggests that port charges per DWT for cellular containerships decrease as ship size (in DWT) increases. However, port charges per number of TEU movements per day in port for cellular containerships may either increase or decrease with respect to increases in the number of TEU movements per day and ship size.

The optimal size for cellular containerships is responsive to changes in the distance of a ship's round trip, container handling efficiency at ports of call, and the concentration of containers at these ports. If the distance of a round trip route increases, the optimal size of cellular containerships serving this route can be expected to increase. If the container handling efficiency at ports of call improve, the optimal size of cellular containerships calling at these ports can be expected to increase. If load centering—the concentration of containers at ports of call—increases, the optimal size of cellular containerships calling at these ports can be expected to increase.

Notes to Chapter 5

1. For further discussion of semicontainer ships, see S. Gilman, *The Competitive Dynamics of Container Shipping* (Aldershot, England: Gower Publishing, 1983), p. 19.

2. For further discussion of shipborne ramps of RO-ROs, see S. Gilman, *Ship Choice in the Container Age* (Surrey, England: McMillian, 1980), p. 119.

3. For further discussion of these capacity measures, see J. O. Jansson and D. Shneerson, "Economies of Scale of General Cargo Ships," *Review of Economics and Statistics* 6(1978):287–293; J. O. Jansson and D. Shneerson, "The Optimal Ship Size," *Journal of Transport Economics and Policy* 16(1982):217–238; and J. O. Jansson and D. Shneerson, *Liner Shipping Economics* (London: Chapman and Hall, 1987).

4. W. K. Talley, "Optimal Containership Size, Load Center Ports and Route Distances," unpublished paper (1989).

5. Jansson and Shneerson, *Liner Shipping*, pp. 126–127.

6. T. Thorburn, *Supply and Demand of Water Transport* (Stockholm: EFI, 1960).

7. The product of holding capacity (TEU) and hauling speed (miles/day) equals hauling capacity (TEU-miles/day).

8. Hauling capacity (HD) equal to the product of holding capacity (HO) and hauling speed (SP) may alternatively be expressed as $\ln HD = \ln HO + \ln SP$. The elasticity of holding capacity (Eho) with respect to ship size (S) may be expressed as $Eho = \ln HO / \ln S$ or $\ln HO = Eho * \ln S$ and the elasticity of hauling speed (Esp) with respect to ship size may be expressed as $Esp = \ln SP / \ln S$ or $\ln SP = Esp * \ln S$. Thus $\ln HD = Eho * \ln S + Esp * \ln S$ or $\ln HD = (Eho + Esp)\ln S$. Thus the elasticity of hauling capacity with respect to ship size equals (Eho + Esp).

9. Talley, "Optimal Containership."

10. W. K. Talley, "A Short-Run Cost Analysis of Ocean Containerships," *The Logistics and Transportation Review* 22:(1986):131–139.

11. Unit sea vessel cost curve EK may be interpreted as a short-run unit cost curve, since size of ship is held constant.

12. Talley, "Short-Run Cost Analysis."

13. The unit sea vessel cost curve in Figure 5.5 may be interpreted as a long-run unit cost curve, since size of ship is varying.

14. Gilman, *Competitive Dynamics* and *Ship Choice*.

15. *Ibid*.

16. Talley, "Optimal Containership."

17. A discussion of load-center ports is found in Talley, "Optimal Containership"; M. L. Chadwin, ed., *Proceedings: The Shipping Act of 1984: Evaluating Its Impact: A Conference Sponsored by the Federal Maritime Commission and Old Dominion University, Norfolk, Virginia, June 12–13, 1986* (Norfolk: Virginia Center for World Trade, 1986); B. E. Marti, "The Evolution of Pacific Basin Load Centres," *Maritime Policy and Management* 15:57–66 (1988); A. Kamada, "Future Development of Container Load Center and Their Associated Inland Transport Systems in Japan," *Future Challenges in Asian Pacific Shipping: Asian Seatransport Conference* (1989).

18. Talley, "Optimal Containership."

19. For further discussion of containerships and containership service, see R. Pearson and J. Fossey, *World Deep-Sea Container Shipping* (Aldershot, England: Gower Publishing Co., 1983); L. C. Kendall, *The Business of Shipping* 4th ed. (Centreville, MD: Cornell Maritime Press, 1983); J. L. Eyre, "The Containerships of 1999," *Maritime Policy and Management* 16 (1989):133–145.

Chapter 6
NETWORKS, INTERMODALISM, AND CONTAINERSHIP SERVICE

In this chapter we continue our discussion of the containership. The focus of the discussion is the interport network, which is a port-connecting route system over which container service is provided—port-to-port container movements. Two types of interport networks are considered—containership and intermodal. A containership interport network is defined as an interport network over which only containerships move containers. In an intermodal interport network, both containerships and other transportation modes are used. We also discuss selection criteria and determinants of interport networks.

NETWORKS

In Chapter 5, we note that containerships exhibit cost economies of ship size at sea but sometimes exhibit cost diseconomies of ship size in port. If so, a containership line has to weigh the cost economies of a larger-size containership at sea against its diseconomies in port in determining the size of ship to provide service over the selected interport network. Determinants of the selected network include the amount of cargo at ports of call, pricing policies of containership lines, container handling efficiency at ports of call, and distances between ports of call. The selected network may either be a containership or an intermodal network.

Containership Networks

Containership networks may be described as origin-to-destination or mainline. We define an origin-to-destination containership network (ORIDES) as a network over which the same containership transports cargo from its origin port through the network to its destination port. A mainline containership network (MAIN) is defined as a network over which the same containership does not necessarily provide containership service for cargo from its origin port to its destinction port. For example, a MAIN network may have a connecting feeder network, where smaller containerships are used to feed cargo into or out the port where the two networks connect. Further, MAIN networks may have transshipment points where cargo is transferred from a containership of one MAIN network to a containership of another MAIN network. In other words, the MAIN networks intersect at the transshipment port. In contrast, an ORIDES network is one for which there are no connecting feeder networks nor transshipment port centers.

In general, two types of ORIDES networks have been utilized by vessel operators— constant frequency and variable frequency. A constant frequency ORIDES network is one over which a containership calls at all of the ports on the network the same number of times (generally once) on a given round trip. A variable frequency ORIDES network is one over which a containership may call on some ports more than others on the same round trip.

A constant frequency ORIDES network is depicted in Figure 6.1. The figure considers two separate ranges of ports, with ports A, B, and C being on one range and ports C, D, and E on the other. A containership calls once at each port on each round trip. If amounts of cargo to be exchanged at individual ports are relatively small, a small ship will likely be used. Thus economies of ship size at sea will be sacrificed to achieve broad market coverage. Also, as the number of port calls increases, ship schedule adherence (and thus the reliability of service) is more likely to be adversely affected.

A variable frequency ORIDES network is depicted in Figure 6.2. A containership calls once at each port on each round trip, except for port A, where it calls twice. The containership service shown in Figure 6.2 achieves broad market coverage while providing more frequent service at ports where there are higher volumes of containers.

MAIN networks may generally be described as feeder or transshipment networks (or a combination of the two). A feeder MAIN network is depicted in Figure 6.3. The MAIN network DEFB has a connecting feeder network Djk, feedering container cargo into and out of D, a port common to both networks. A relatively small vessel serves feeder network Djk and a relatively large one operates along MAIN network DEFB. The containership service in Figure 6.3 achieves broad market coverage in one region while allowing the ship line to take advantage of cost economies of ship size at sea on the MAIN network. By limiting the number of port calls on the MAIN network, a containership's cost diseconomies of ship size in port (if they exist) are reduced and frequency of service is enhanced. With the increased concentration of containers at port

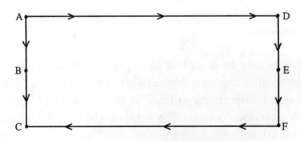

Figure 6.1. Constant frequency ORIDES network.

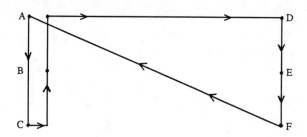

Figure 6.2. Variable frequency ORIDES network.

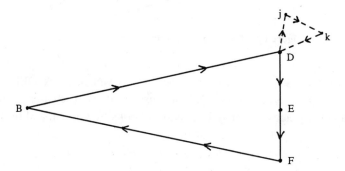

Figure 6.3. Feeder main network.

D due to the feeder service, the line can take advantage of a containership's cost economies of ship utilization (or load factor) in port at port D.

In Figure 6.4, transshipment MAIN networks are depicted. MAIN networks ABCE and GHIE have in common port E, a transshipment center. At port E, containers going to ports on one network are transferred from vessels operating on the second network to ships serving the first network. The advantages are similar to feeder MAIN networks. In addition, both vessel operators are able to take advantage of cost economies of ship size at sea, since smaller containerships are not being used as in the case of feeder service.

In Figure 6.5, a combination of feeder and transshipment MAIN networks has been created by merging networks depicted in Figures 6.3 and 6.4. MAIN networks DEFB

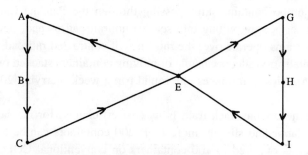

Figure 6.4. Transshipment main networks.

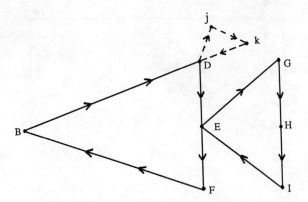

Figure 6.5. Feeder and transshipment main networks.

and GHIE have a transshipment center at port E. MAIN network DEFB has a connecting feeder network Djk at its port D. A discussion of specific cases of these general types of containership networks may be found in books by Gilman and by Pearson and Fossey.[1]

Intermodal Networks

As stated previously, intermodal networks are networks over which both containerships and alternative transportation modes are used to transport containers. The alternative modes may be either land-based (for example, double-stack trains) or water-based (for example, barges). The land-based transportation service between two ports has been referred to as either "landbridging" or "mini-landbridging." As noted in Chapter 1, landbridging refers to the movement of cargo across a body of land between two ocean legs (for example, eastward by ship across the Pacific to the West Coast of the United States, by train to a U.S. East Coast port, and across the Atlantic by ship to Europe). Mini-landbridging refers to cargo movement that crosses one ocean by ship and then crosses a body of land but ends at a port on another ocean.[2]

An example of land-based bridging service initiated in the United States is double-stack container rail service. In April 1984 the containership line, the American President Lines (APL), in response to the round-the-world service competition of United States Lines (utilizing jumbo containerships passing through the Panama Canal) committed itself to bridging service by putting into service nonrailroad-owned, double-stack container trains. The trains operate over the rail lines of contracted railroads. Double-stack trains consist of platform railcars capable of moving containers stacked two high. APL's Chicago to Los Angeles train makes one round trip a week, carrying 200 containers in each direction.

The appeal of the double-stack train is its cost efficiency; for the same locomotive power, the same labor and slightly more fuel, 200 containers can be transported on a double-stack train as opposed to 100 containers on conventional COFC (container on flat car) trains. Huneke, Lane, and Benforado conclude that the United States Lines' round-the-world containership service (passing through the Panama Canal) had a cost advantage over double-stack train service for eastbound Pacific Coast cargo to New York City but not for eastbound Pacific Coast cargo to Chicago.[3] Even if double-stack train service does not have a cost advantage over containership service, it offers significant savings in transit time (and thus inventory cost savings) for shippers.[4]

NETWORK SELECTION

How should containership lines go about selecting their (interport) networks? The criteria should consider the trade-offs among all possible containership and intermodal networks in the geographical regions where they want to operate. In the following discussion, we assume the vessel operators seek to maximize profits in the selection of networks. Based on this assumption, a criterion is presented for selecting a network from among all the possibilities in a given region.

A Containership Line's Perspective

Suppose all the possible networks are just containership networks (only containerships are to be used). Suppose as well that these possible networks consist of only four ports.[5] Since the preferred front-haul trip network may differ from the preferred back-haul network for these ports, the selection criterion will consider all possible networks for which cargo is moving in the same direction. For the cargo moving in the opposite direction, the selection criterion will have to be applied a second time.

Suppose there are three ports (A, B, and C) along a given coastline where container cargo destined for port D originates. Four possible networks are diagrammed in Figures 6.6, 6.7, 6.8, and 6.9. Figure 6.6 depicts a network where three different vessels are used to provide direct service from ports A, C, and B to port D. Figure 6.7 depicts a network where a containership provides direct service from port B to port D, but cargo going from port A to port D goes via port C along with cargo from port C itself. The movement of port A's cargo may be in one ship or two (with one vessel serving as a feeder from port A to port C). In Figure 6.8, cargo going from port B to port D goes via port C, and in Figure 6.9, cargo going from both port B and port A travels via port C. For the sake of simplicity, other possible networks are not considered (for example, networks based upon routings B to C to A to D or A to C to B to D). Further, we assume that the appropriate size ship (or ships) for the amounts of cargo are used, namely, those that minimize the cost of providing the service.

Under the assumption that a containership line seeks to maximize profits (Π) and that Network Four (see Fig. 6.9) is the selected network,[6] it thus follows that Network Four's profits are greater than the profits of each alternative network or:

$$\Pi_{ACBCD} > \Pi_{AD} + \Pi_{BD} + \Pi_{CD} \qquad (6.1)$$

$$\Pi_{ACBCD} > \Pi_{ACD} + \Pi_{BD} \qquad (6.2)$$

$$\Pi_{ACBCD} > \Pi_{BCD} + \Pi_{AD} \qquad (6.3)$$

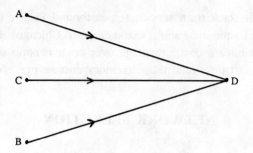

Figure 6.6. Network one.

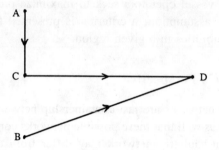

Figure 6.7. Network two.

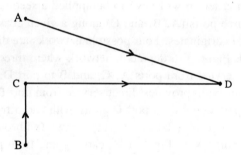

Figure 6.8. Network three.

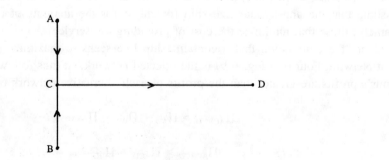

Figure 6.9. Network four

Equation (6.1) states that the profit of Network Four following the route structure ACBCD is greater than Network One's (see Fig. 6.6) profit—the sum of the profits of routes AD, BD, and CD. Equation (6.2) states that Network Four's profit is greater than Network Two's (see Fig. 6.7) profit—the sum of the profits of routes ACD and BD. Equation (6.3) states that Network Four's profit is greater than Network Three's (see Fig. 6.8) profit—the sum of the profits of routes BCD and AD.

Based upon the above assumptions, we can now deduce the following criterion for selecting a containership network: the containership network that should be chosen is the one for which the costs saved are greater than the revenue lost, when compared to all possible alternative networks. The derivation of this selection criterion appears in the appendix to this chapter.[7] In the appendix, the cost savings and loss in revenue in utilizing the selected containership network rather than an alternative containership network are formulated.[8]

Cost savings for the containership line may arise from feedering cargo from one port to another and then combining this cargo with cargo at the latter port for transport on a larger containership to the destination port (rather than providing direct service from origin to destination ports). However, with this cost savings, there will be a loss in revenue. Feeder containership service generally is a poorer quality of service than direct containership service from the standpoint of tne shipper, since its transit time usually is greater.

In addition to utilizing only containerships to move containers over networks, a shipping line may also utilize both containerships and other transportation modes. If services of these modes are under the control of shipping line, then the line may make use of the above selection criterion for selecting an intermodal network by considering the cost and revenue functions for the services to be provided by these modes. The control of services provided by other transportation modes by a shipping line may take the form of ownership or a contract for service by another firm.[9]

A National Perspective

The above discussion presents a criterion that may be used by a containership line for selecting networks. However, the selected networks and thus the ports to be called upon and the frequency of port calls may not be those that are desirable from the standpoint of a country's national economy. For example, the shipping line's objective of maximizing profits may be in conflict with a country's national objective of increasing throughput through certain ports to stimulate economic growth (and thus employment) in the region. A methodology for addressing this issue is found in a report by the Australian Bureau of Transport Economics (BTE).[10] The BTE report sought to establish whether changes in the existing pattern of ocean transport networks serving Australia could produce greater economic benefits not only for the nation, but also for the affected parties. The BTE report presents a methodology for determining the costs as well as the cost savings related to an additional port call by a ship diverted from its existing route. The methodology assumes that the country's ports have adequate capacity to meet the demands placed upon it. Further, only those costs that can be varied in the short-run are included in the calculations.

The report considering four broad categories of costs (and thus potential cost savings): (1) ship-related costs, (2) port-tug and pilotage-related costs, (3) terminal costs, and (4) land transport costs. Ship-related costs are those costs for an additional port call incurred by a diverted ship. These ship-related costs include additional fuel consumed at sea and during pilotage at the additional port call, and marginal repairs and maintenance caused by the additional call. Port-tug and pilotage-related costs are fuel and marginal repair and maintenance costs incurred by tug and pilot boats for the additional port call by the diverted ship. Terminal costs include costs incurred by a marine terminal with respect to the additional port call. Land transport costs include costs incurred by such land transport providers as railroads and truck lines with respect to the additional call.

The BTE report evaluated ten alternative port call strategies. Each strategy involved a ship being diverted from its existing network to make an additional port call at either Adelaide or Brisbane and then returning to its original itinerary. Where cost savings were found, further investigation was made to establish the number of containers needed on each visit so that the carrier would at least break even on the extra call. Not surprisingly, the report found that the longer the diversion, the greater the number of containers needed to make the additional calls worthwhile.[11]

NETWORK DETERMINANTS

In the above discussion, the criterion for selecting networks from all possible containership and intermodal networks by a containership line involves comparing the cost savings versus the loss in revenue in choosing a given network over an alternative network. What are the factors that affect the revenue received and cost incurred by containership lines in the provision of service in a network? Some of the more important ones include port consignment size (cargo consigned to a ship per port call), liner pricing policies, convexity ratios, ship time in port, and the extent of economic regulation (or deregulation) of ocean and intermodal transportation. Let us look at each.

Port Consignment Size

As discussed in Chapter 5, there is a positive relationship between the optimal size of a containership and the concentration of containers at its ports of call. A larger consignment size allows a containership to take advantage of its cost economies of ship utilization (or load factor) in port. Further, larger consignment sizes provide an opportunity for containerships to call at fewer ports, thus increasing their days at sea. Hence, larger consignment sizes provide an opportunity for containership lines to take advantage of the cost economies of containership utilization (or increasing the load factor of a given size ship) at sea as well as the cost economies of containership size (or utilizing a larger size ship) at sea.

Ports where there are naturally relatively high concentrations of containers or where cargo is concentrated through feeder services are often referred to as load-center ports.[12] In 1988 the world's largest load-center port was Hong Kong, which handled a volume of 4 million TEUs; Singapore was ranked second with a volume of 3 million TEUs. On the U.S. Atlantic coast, New York is the North Atlantic load-center port; the Port of Hampton Roads (Norfolk) and Baltimore are competing to be the mid-Atlantic load-center port; and Charleston appears to have displaced Savannah as the South Atlantic load-center port.[13]

Networks consisting of load-center ports are often utilized by containership lines to provide round-the-world containership service. Evergreen Line provides two-way (or bidirectional) round-the-world service with a round trip transit time of 77 days. The 21 ports of call on Evergreen's eastbound round-the-world service are: Singapore, Hong Kong, Kaohsiung, Keelung, Pusan, Hakata, Osaka, Nagoya, Shimizu, Tokyo, Los Angeles, Charleston, Baltimore, New York, Le Havre, Antwerp, Rotterdam, Felixstowe, Hamburg, Colombo, and Port Kelang. The ports of call on Evergreen's westbound round-the-world service include: Tokyo, Nagoya, Osaka, Pusan, Keelung, Kaohsiung, Hong Kong, Singapore, Colombo, Hamburg, Felixstowe, Rotterdam, Antwerp, Le Havre, New York, Norfolk, Charleston, Kingston and Los Angeles. As noted in Chapter 1, other containership lines providing round-the-world service in a single vessel include Senator Line and Nedlloyd.[14]

Liner Pricing Policies

Prior to containerization, liner pricing structures generally incorporated the practice of equalization.[15] In equalization pricing, the freight rate for a given type of cargo is the same from any main port in a range on one end of a vessel's route to any main port in a range on the other end. Since the shipper is responsible for inland transport costs to and from ports, the shipper could minimize his total transport costs (ocean and inland) by shipping his cargo out of the nearest port. Thus ports developed natural hinterlands, and a ship line needed to call at each port if it desired to obtain the cargo from each hinterland. Equalization pricing was, thus, a major factor contributing to the extensive multiport itineraries and the duplication of port calls by liner companies prior to containerization.[16]

With the advent of containerization, different route structures (and networks) were needed for containership lines to take advantage of the cost economies of containership size at sea. Further, relatively large port consignment sizes were needed to secure these cost economies. It was no longer desirable for a shipping line to have extensive multiport itineraries. The liner pricing structure that developed to promote the cost economies of containerships was absorption pricing.

Absorption pricing is a liner pricing structure under which shippers are charged a door-to-door rate independent of port choice. The shipper is charged for inland transport as though the cargo were going to its nearest port, irrespective of the port from which it actually sails. Thus a shipping line in theory need not make a direct call to a port if inland transport can be used as a substitute for ship diversion. Absorption pricing dissipates the hinterlands of ports that developed under conventional liner pricing. Under absorption pricing, the choice of port has shifted from the shipper to the shipping line, and the decision to call at a port will hinge on the economic trade-offs between diverting a mainline vessel, using a feedership or using alternative land transport modes.[17]

Absorption pricing provides support for the argument that ship lines should make one port call per country. However, as a practical matter, this has not been the normal practice. Containership lines have adopted multiport itineraries for some countries for several reasons. First, substitution of land transport for containership service is not necessarily cheaper. Second, ship lines have an incentive to call at ports that have higher valued (though less concentrated) cargo for which they can charge relatively higher freight rates. Third, fearful of losing cargo to a competitor, a ship line may be unwilling to take the risk of calling at only one port per country.[18]

Other Determinants

In addition to port consignment size and liner pricing policies, other important determinants of networks for containership lines are the convexity ratio, ship time in port, and the extent of regulation (or deregulation) of ocean and intermodal transportation. The convexity ratio for ocean transportation is the ratio between maritime distance saved (or incurred) and inland distances thereby incurred (or saved). Once a network is defined, a call at a neighboring port may add little additional maritime distance while perhaps achieving significant savings in inland distances. Alternatively, incurring inland distances as in landbridge service may simultaneously save significant maritime distances. If the latter is true, an intermodal network will likely be used rather than a containership network—if inland modes are available, reliable, and cost efficient.

In order to take advantage of containerships' cost economies of ship size at sea, shipping lines will seek to minimize the amount of time their containerships are in port, as noted earlier. Hence, lines will seek to use those ports offering relatively faster ship turnround times.[19] Further, shipping lines are likely to be more sensitive to ship turnaround times than to port charges in port selection, since port charges are a relatively small proportion of the total cost of containership service. Where competition among neighboring ports exists, however, ports may seek to attract containership calls from their neighbors by lowering port charges.[20]

Another important determinant of networks for containership lines is the extent of economic regulation (or deregulation) of ocean and intermodal transportation. If regulation limits intermodal service combining ocean and land carriers, a containership line's ability to form intermodal networks will obviously be limited. Without freedom in pricing, the formation of intermodal networks will also be constrained. The type of route systems (multiport versus limited-port) for containership networks will be affected as well.

To offer insight into the impact that economic deregulation can have on ocean container transportation, the following section discusses the impact of deregulation of ocean and land transportation in the United States on ocean container transportation.

DEREGULATION AND OCEAN CONTAINER TRANSPORTATION: THE U.S. CASE

The U.S. ocean container transportation system has undergone major changes in recent years resulting from economic deregulation in the truck, railroad, and ocean transportation industries. The Motor Carrier Act of 1980 deregulated U.S. truck carriers by granting them greater ratemaking freedom, increased opportunities for new carriers to enter the industry, and created new opportunities for existing carriers to expand their services. As a consequence, truck carriers have contracted with containership lines to provide door-to-door, international intermodal service. Similarly, the Staggers Act of 1980 deregulated U.S. railroads by granting these carriers greater ratemaking freedom, the freedom to enter into service contracts with shippers, and the flexibility to market their intermodal services.[21] The Shipping Act of 1984 deregulated the U.S. ocean transportation industry by reducing the regulatory burden of ocean carriers. The act authorized service contracts and intermodal rates; permitted independent action on rates and service by conference carrier members; expedited the review process of agreements by the U.S. regulatory commission, the Federal Maritime Commission; and broadened the antitrust immunity of the collective actions of ocean carriers.[22]

Prior to deregulation, U.S. ocean ports were considered to have natural hinterlands. All shipment points that were closer to a given port than to any other port comprised the natural hinterland of the port. Under deregulation, however, the area (in square miles) of these natural hinterlands has greatly decreased. "Port proximity and inland cost are no longer the deciding factors in the international shipper's port selection criteria."[23] Under deregulation, ocean carriers have a greater say in port choice, especially when they enter into contracts with inland carriers to provide through-service. Under such contracts, the transportation cost of intermodal service is not necessarily minimized by the transport of cargo to its nearest port. Further, with increased intermodal competition under deregulation, shippers are able to select carriers that provide services at the lowest logistics costs (such as inventory, warehousing, and transportation) rather than at the lowest transportation cost. Consequently, carrier service rather than the distance to the closest port is now the relevant determinant of port choice by a shipper.

Under deregulation, the competition between landbridge and all-water services has intensified. Utilizing double-stack train operations and intensive marketing efforts, shipping lines such as APL and SeaLand quickly established their dominant position in the intermodal transportation market. The success of APL and SeaLand has resulted in other major containership lines such as Maersk and Evergreen following their methods. Double-stack operations are geographically concentrated between the U.S. West Coast and the Midwest. The two primary reasons for this occurrence are: (1) the prominence of U.S. trade with the Far East, (2) the long distance between the West Coast and the Midwest for which rail service holds a cost advantage over truck service, and (3) the relative absence of tunnels and underpasses as compared to the East Coast.

As mentioned in Chapter 1, a problem that has arisen with double-stack train operations is the imbalances in the flows of international cargo. Operators of double-stack train service are faced with the prospect of either having significant numbers of empty containers or having to obtain domestic cargo for backhauls. Many containership operating double-stack service have chosen the latter. As a consequence, domestic cargo accounted for over 90 percent of the cargo on APL's westbound double-stack trains in 1987.[24]

In addition to the marketing strategies of carriers, the marketing strategies of ports have also been affected by deregulation, as discussed in Chapter 1. Ports are now in competition with one another to attract containership lines to call at their ports. They are under pressure to improve their efficiency in the handling of containers and to expand or alter their capabilities structures in order to succeed in this new competitive environment. In some cases, this pressure has manifested itself in ports seeking to become load-center ports as well as seeking to attract double-stack train services.[25]

SUMMARY

Networks for container service may be classified as containership or intermodal networks. Containership networks are interport networks over which only containerships are used to provide container transport service between ports. Intermodal networks are interport networks over which both containerships and alternative modes are used. Containership networks, in turn, may be described as origin-to-destination or mainline. The same containership transports cargo from its origin port to its destination port over a origin-to-destination network; the frequency of port calls may be constant or variable for a given round-trip route. A mainline containership network does not necessarily use the same containership to transport cargo from its origin port through the network to its destination port. A mainline network may have an adjacent feeder network and/or a transshipment port.

The criteria used to select networks from among all possible networks should consider the trade-offs, for example, the cost savings to the carrier versus the revenue loss by utilizing one network rather than another. From the perspective of a country's national economy, the basis for the trade-off analysis may be economic benefits versus costs incurred. For the containership line some of the more important determinants of network selection include port consignment size, liner pricing policies, convexity ratios, ship time in port, and the extent of economic regulation of ocean and intermodal transportation.

APPENDIX

Containership Network Selection Criterion

The containership network that should be chosen is the one for which the costs saved are greater than the revenue lost, when compared to all possible alternative networks.

The profit terms in equations (6.1), (6.2) and (6.3) may be defined as follows:

$$\Pi_{AD} = P_{AD}^A Q_{AD}^A - C(Q_{AD}^A, O, O) \qquad (a)$$

where, Q_{AD}^A is the amount of cargo originating at port A following the route AD; P_{AD}^A is the transport price per unit of cargo Q_{AD}^A; and $C(Q_{AD}^A, O, O)$ is the cost incurred by the containership line in transporting cargo Q_{AD}^A.

$$\Pi_{BD} = P_{BD}^B Q_{BD}^B - C(O, Q_{BD}^B, O) \qquad (b)$$

where, Q_{BD}^B is the amount of cargo originating at port B following the route BD; P_{BD}^B is the transport price per unit of cargo Q_{BD}^B; and $C(O, Q_{BD}^B, O)$ is the cost incurred by the containership line in transporting cargo Q_{BD}^B.

$$\Pi_{CD} = P_{CD}^C Q_{CD}^C - C(O, O, Q_{CD}^C) \qquad (c)$$

where, Q_{CD}^C is the amount of cargo originating at port C following the route CD; P_{CD}^C is the transport price per unit of cargo Q_{CD}^C; and $C(O, O, Q_{CD}^C)$ is the cost incurred by the containership line in transporting cargo Q_{CD}^C.

$$\Pi_{ACD} = P_{ACD}^A Q_{ACD}^A + P_{CD}^C Q_{CD}^C - C(Q_{AC}^A, O, O) - C(Q_{CD}^A, O, Q_{CD}^C) \qquad (d)$$

where, Q_{ACD}^A is the amount of cargo originating at port A following the route ACD; P_{ACD}^A is the transport price per unit of cargo Q_{ACD}^A; $C(Q_{AC}^A, O, O)$ is the cost incurred by the containership line in transporting cargo Q_{AC}^A; Q_{AC}^A is the amount of cargo originating at port A following the route AC; $C(Q_{CD}^A, O, Q_{CD}^C)$ is the cost incurred by the containership line in jointly transporting cargos Q_{CD}^A and Q_{CD}^C; and Q_{CD}^A is the amount of cargo originating at port A following the route CD. It follows that $Q_{ACD}^A = Q_{AC}^A = Q_{CD}^A$.

$$\Pi_{BCD} = P_{BCD}^B Q_{BCD}^B + P_{CD}^C Q_{CD}^C - C(O, Q_{BC}^B, O) - C(O, Q_{CD}^B, Q_{CD}^C) \qquad (e)$$

where, Q_{BCD}^B is the amount of cargo originating at port B following the route BCD; P_{BCD}^B is the transport price per unit of cargo Q_{BCD}^B; $C(O, Q_{BC}^B, O)$ is the cost incurred by the containership line in transporting cargo Q_{BC}^B; Q_{BC}^B is the amount of cargo originating at port B following the route BC; $C(O, Q_{CD}^B, Q_{CD}^C)$ is the cost incurred by the containership line in jointly transporting cargos Q_{CD}^B and Q_{CD}^C; and Q_{CD}^B is the amount of cargo originating at port B following the route CD. It follows that $Q_{BCD}^B = Q_{BC}^B = Q_{CD}^B$.

$$\Pi_{ACBCD} = P_{ACD}^A Q_{ACD}^A + P_{BCD}^B Q_{BCD}^B + P_{CD}^C Q_{CD}^C - C(Q_{AC}^A, O, O)$$
$$- C(O, Q_{BC}^B, O) - C(Q_{CD}^A, Q_{CD}^B, Q_{CD}^C) \qquad (f)$$

where, $C(Q_{CD}^A, Q_{CD}^B, Q_{CD}^C)$ is the cost incurred by the containership line in jointly transporting cargos Q_{CD}^A, Q_{CD}^B and Q_{CD}^C.

Substituting equations (a), (b), (c) and (f) into equation (6.1) and rewriting, we obtain:

$$[C(Q_{AD}^A, O, O) - C(Q_{AC}^A, O, O)] + [C(O, Q_{BD}^B, O) - C(O, Q_{BC}^B, O)]$$
$$+ [C(O, O, Q_{CD}^C) - C(Q_{CD}^A, Q_{CD}^B, Q_{CD}^C)] > [P_{AD}^A Q_{AD}^A - P_{ACD}^A Q_{ACD}^A]$$
$$+ [P_{BD}^B Q_{BD}^B - P_{BCD}^B Q_{BCD}^B] \tag{1'}$$

The expression to the left side of the inequality sign is interpreted as the cost savings to the containership line in using Network Four rather than Network One. The expression to the right side of the inequality sign is interpreted as the loss in revenue to the containership line in using Network Four rather than Network One.

Substituting equations (b), (d) and (f) into equation (6.2) and rewriting, we obtain:

$$[C(O, Q_{BD}^B, O) - C(O, Q_{BC}^B, O)] + [C(A_{CD}^A, O, Q_{CD}^C) - C(Q_{CD}^A, Q_{CD}^B, Q_{CD}^C)]$$
$$> P_{BD}^B Q_{BD}^B - P_{BCD}^B Q_{BCD}^B \tag{2'}$$

The expression to the left side of the inequality sign is interpreted as the cost savings to the containership line in using Network Four rather than Network Two. The expression to the right side is interpreted as the loss in revenue in using Network Four rather than Network Two.

Substituting equations (a), (e) and (f) into equation (6.3) and rewriting, we obtain:

$$[C(Q_{AD}^A, O, O) - C(Q_{AC}^A, O, O)] + [C(O, Q_{CD}^B, Q_{CD}^C) - C(Q_{CD}^A, Q_{CD}^B, Q_{CD}^C)]$$
$$> P_{AD}^A Q_{AD}^A - P_{ACD}^A Q_{ACD}^A \tag{3'}$$

The expression to the left side of the inequality sign is interpreted as the cost savings to the containership line in using Network Four rather than Network Three. The expression to the right side is interpreted as the loss in revenue in using Network Four rather than Network Three. With Network Four being the chosen network and since its cost savings when compared to each possible alternative network is greater than the lost in revenue, it thus follows that we have proven the criterion.

Notes to Chapter 6

1. S. Gilman, *Container Logistics and Terminal Design* (Washington, DC: International Bank for Reconstruction and Development, 1981), pp. 65–78; R. Pearson and J. Fossey, *World Deep-Sea Container Shipping* (Aldershot, England: Gower Publishing Co., 1983), pp. 96–113.

2. Cargo movement that crosses one ocean by ship and then proceeds by rail, for example, to an inland location (rather than port to port by land) is referred to as "microbridging."

3. W. F. Huneke, L. L. Lane, and D. J. Benforado, "A Competitive Analysis of United States Lines Round-the-World Service," *Annual Proceedings of the Transportation Research Forum* 27 (1986):22–29. For further discussion of double-stack train service in the United States, see L. T. Thuong, "From Piggyback to Double-Stack Intermodalism," *Maritime Policy and Management* 16 (1989):69–81.

4. Discussions of inventory costs in ocean shipping are found in J. O. Jansson and D. Shneerson, "A Model of Scheduled Liner Freight Services: Balancing Inventory Cost Against Shipowners' Costs," *The Logistics and Transportation Review* 21 (1985):195–215, and J. A.

Pope and W. K. Talley, "Inventory Costs and Optimal Ship Size," *The Logistics and Transportation Review* 24 (1988):107–120. An alternative water transport service that has been utilized in intermodal networks is barge service—container-on-barge and shipborne-barge. In container-on-barge service, conventional containers are transported by barges for transfer to and from containerships. In shipborne-barge service, the barge serves as a floating container and the barge itself is transferred to and from specially designed ships. For discussions of such service, see J. G. Crew and K. H. Horn, "Assessment of Container-on-Barge Service on the Mississippi River System," *Journal of the Transportation Research Forum* 28 (1987):92–96 and M. H. Sonstegaard, "World Standards for Shipborne Barges," *Transportation Research* 21A (1987):139–144.

5. Only four ports are considered in order to simplify the analysis. A linear increase in the number of ports causes the number of possible networks to grow geometrically.

6. Since it makes no difference which network is designated as the selected one, network four was arbitrarily selected.

7. This selection criterion and its proof are found in W. K. Talley, "Load Center Ports and Interport Networks for Containerships," an unpublished paper (1989). An indifference criterion for a containership line to be indifferent among possible networks is also found in this paper.

8. A number of other approaches are available for selecting the optimal network. An example is what has historically been referred to as the "traveling salesman" problem. In the maritime version of this problem, a ship line operator would be faced with serving a number of ports in a network. He would need to determine the optimal route that would result in a containership calling at each port once and returning to its starting point without retracing any routes or visiting any port more than once. The criterion for selecting the optimal route could be any quantifiable value—distance or cost, for example. The result of solving the problem would be the route through the network that would minimize the distance or cost or whatever other criterion the operator selected. There are a number of methods for solving this problem, but they have one important characteristic in common—they do not require the evaluation of all possible routings in order to find the best routing. A network consisting of only five ports, for example, has 120 possible routings through the network. A network consisting of 10 ports has 3,628,800 possible routings. Techniques that require one to consider all the possible routings soon bog down and become impossible to use. Network algorithms, such as those used to solve the "traveling salesman" problem, work efficiently by ignoring routings that are obviously inefficient or nonoptimal. Routings may be determined quickly for fairly large networks using common computer routines. For further discussion of network algorithms, see L. R. Ford and D. R. Fulkerson, *Flows in Networks* (Princeton: Princeton University Press, 1962).

9. In addition to selecting intermodal interport networks utilizing landbridging and minilandbridging, the selection criterion can also be applied to noninterport networks where microbridging service is utilized (assuming such service is under the control of the shipping line). Specifically, revenue and cost functions related to transport service to and from ports would be adjusted to include the revenue and cost related to inland cargo movements to and from ports. Such a criterion could consider the trade-off in port cost savings of a shipping line calling at fewer ports (or load-center ports) versus the increase in inland transportation costs. If the microbridging service is not under the control of the shipping line, the network (or networks) selected by the shipping line will depend upon such factors as the level of competition that exists among neighboring ports, inland carriers, and shipping lines; the rates and service provided by ports and inland carriers; and the amounts of cargo at various ports.

10. Bureau of Transport Economics, *Cargo Centralisation in the Overseas Liner Trades* (Canberra, Australia: Research Report No. 52, 1982).

11. In addition to considering containership and intermodal network selection from the perspective of the shipping line and a nation, it can also be considered from the perspective of the

shipper. The shipper may incur cost savings, for example, from shipping cargo in the first-sailing vessel even if the transit time is longer, since the vessel serves as a free-storage warehouse.

12. The current literature lacks a precise definition of load-center ports. Alternative definitions are found in B. E. Marti, "The Evolution of Pacific Basin Load Centres," *Maritime Policy and Management* 15 (1988):57–66; Talley, "Load Center Ports"; and W. K. Talley, "Optimal Containership Size, Load Center Ports and Route Distances," an unpublished paper (1989).

13. B. Vail, "Concept of Load Centers Has Lost Some of Its Shine," *Journal of Commerce* (June 13, 1989):26C.

14. Prior to bankruptcy, United States Lines (USL) provided round-the-world containership service over a network that included 14 load-center ports and 33 feeder ports. The USL utilized its jumbo containerships to transport cargo between load-center ports and foreign-flag feeder ships to move cargo between feeder ports and load-centers. Foreign-flag vessels were utilized, since they were less costly to operate than U.S. flag vessels. A discussion of the rationale for round-the-world service is found in K. Y. Hwang, "Why Round-The-World?" *Future Challenges in Asian Pacific Shipping: Asian Seatransport Conference* (1989).

15. Discussions of liner pricing are found in T. D. Heaver, "Trans-Pacific Trade, Liner Shipping and Conference Rates," *The Logistics and Transportation Review* 8 (1972):3–28; T. D. Heaver, "The Structure of Liner Conference Rates," *Journal of Industrial Economics* 21 (1973): 257–265; I. Bryan, "Regression Analysis of Ocean Liner Freight Rates on Some Canadian Export Routes," *Journal of Transport Economics and Policy* 8 (1974):161–173; D. Shneerson, "The Structure of Liner Freight Rates: A Comparative Route Study," *Journal of Transport Economics and Policy* 10 (1976):52–67; E. T. Laing, "The Rationality of Conference Pricing and Output Policies, Part 1," *Maritime Studies and Management* 3 (1975):103–111; E. T. Laing, "The Rationality of Conference Pricing and Output Policies, Part 2," *Maritime Studies and Management* 3 (1976):141–151; and R. Byington and G. Olin, "An Econometric Analysis of Freight Rate Disparities in U.S. Liner Trades," *Applied Economics* 15 (1983):403–407.

16. For further discussion of equalization pricing, see Gilman, *Container Logistics*, pp. 54–55.

17. For further discussion of absorption pricing, see Gilman, *Container Logistics*, pp. 55–56.

18. For further discussion of this rationale, see R. Pearson and J. Fossey, *World*, p. 95. For a discussion of liner pricing under containerization, see W. K. Talley and J. A. Pope, "Determinants of Liner Conference Rates Under Containerization," *International Journal of Transport Economics* 12 (1985):125–155.

19. A discussion of ship turnaround time is found in R. Robinson, "Size of Vessels and Turnround Time: Further Evidence From the Port of Hong Kong," *Journal of Transport Economics and Policy* 12 (1978):161–178, and T. D. Heaver and K. R. Studer, "Ship Size and Turnround Time: Some Empirical Evidence," *Journal of Transport Economics and Policy* 6 (1972):32–50.

20. A discussion of competition among ports is found in D. Bobrovitch, "Decentralized Planning and Competition in a National Multiport System," *Journal of Transport Economics and Policy* 16 (1982):31–42, and W. K. Talley, "The Role of US Ocean Ports in Promoting an Efficient Ocean Transportation System," *Maritime Policy and Management* 15 (1988):147–155.

21. For further discussion of the Motor Carrier Act of 1980 and the Staggers Act of 1980, see W. K. Talley, *Introduction to Transportation* (Cincinnati: South-Western Publishing Co., 1983).

22. For further discussion of the Shipping Act of 1984, see P. M. Donovan, J. C. Godwin, and L. V. Kessler, "The Shipping Act of 1984," *ICC Practitioners' Journal* 51 (1984):463–475; P. A. Friedman and J. A. Devierno, "The Shipping Act of 1984: The Shift from Govern-

ment Regulation to Shipper Regulation," *Journal of Maritime Law and Commerce* 15 (1984):311–351; and M. L. Chadwin, ed., *Proceedings: The Shipping Act of 1984: Evaluating Its Impact: A Conference Sponsored by the Federal Maritime Commission and Old Dominion University, Norfolk, Virginia, June 12–13, 1986* (Norfolk: Virginia Center for World Trade, 1986).

23. J. H. Foggin and G. N. Dicer, "Disappearing Hinterlands: The Impact of the Logistics Concept on Port Competition," *Annual Proceedings of the Transportation Research Forum* 26 (1985):385–390.

24. B. Carey, "How it Stacks up in Chicago." *American Shipper* (1987):39.

25. For a discussion of the impact of the Shipping Act of 1984 on U.S. ports, see W. K. Talley, "The Role of US Ocean Ports in Promoting an Efficient Ocean Transportation System," *Maritime Policy and Management* 15 (1988):147–155.

Chapter 7

OCEAN CONTAINER TRANSPORT IN THE FUTURE

The preceding chapters examine the technique and characteristics of ocean container transport today. But what will they be tomorrow? What trends are likely to be present through the last decade of the twentieth century and beyond? How will the vessels, terminals, and intermodal systems of the future change?

It is always tempting to assume that the immediate future will be an extension of the immediate past. In this case, particularly, it is reasonable to assume that an important part of the future will involve the widening application of existing patterns and the more complete exploitation of existing technologies. In part, this will mean the application in smaller ports and terminals and in developing countries of techniques and technologies that now are utilized only in the most advanced ports and the most developed countries.

The next few years also are likely to see the continuation of trends already in place, in part because some of them involve long-lead time investments that already have been made. Thus one can predict with confidence the continued growth of containerization and intermodalism and, with it, the appearance in the first half of the 1990s of substantially larger (and somewhat faster) containerships as well as terminal equipment that can move and store containers more quickly. Similarly, the rapid spread of sophisticated computerized terminal planning and operating systems seems poised to begin, as does the use of standardized transportation and commercial documents that will be transmitted electronically and accepted wherever significant amounts of commerce occur.

On another level, it seems reasonable to expect continued rationalization of services and concentration among firms in the industry, and the development of giant multi-modal providers of transport services, at least in the major markets. The European Community's "1992" economic integration initiative makes it most likely that Western Europe will follow North America in the development of huge, "full service" inter-modal firms. This evolution probably will involve not only the expansion of land-based transportation systems seaward (like CSX) and the extension inland of ocean carriers (like APC), but also the participation of firms whose origins are in air transport. How-

111

ever, uncertainties are very great, and different political and economic futures could radically alter the accuracy of any forecast almost overnight. As the chief executive officer of one intermodal transportation firm warned his stockholders at the end of 1988.

> ... Factors such as general economic conditions, levels of demand for imports and exports, the relative strength of the U.S. dollar, levels of container capacity deployed in the ... trade, fuel costs, and certain political events could effect the company's future results. These factors are generally not within the control of the company, and the company is unable to predict how these and other economic and competitive conditions will develop or the effect of such conditions on the company's future results.[1]

Even if one assumes that global or major regional military conflicts will be avoided, political and governmental uncertainities abound. Will the PRC and the Soviet Union continue to pursue policies that involve their rapid integration into the mainstream of global commerce? Will the "Europe 1992" initiative quickly yield a truly integrated common market? How much of Western Europe (or even Eastern Europe) will that integration embrace? And will the common commercial and transportation policies that emerge increase protectionism or encourage global as well as regional economic integration? Will the 1990s see the continuation of multilateral agreements improving trade flows worldwide or the development of increasingly separate trading blocs—in North America and in East Asia as well as Europe? Will there be substantial changes in U.S. policy on the regulation of ocean shipping as well as trade and foreign investment? What policy choices will the governments of the major ship-owning and ship-building countries make, for example, between subsidizing these industries and encouraging rationalization through down-sizing, mergers, consortia, and capacity-reducing or capacity-sharing arrangements?

The answers to these questions could become quite different than they appear at this moment. If even a few of them change substantially, the future pace and pattern of containerization and intermodalism could be affected greatly. With these uncertainties in mind, this chapter examines the future of the ocean container carriers, ports and terminals, and, finally, intermodalism.

THE OCEAN CARRIERS

Containerization Continues

During the mid-1980s there was some speculation among transportation experts that containerization had pretty well run its course. The argument was that nearly all of the cargo that was technologically feasible and commercially attractive to containerize had already been containerized. The cargoes that remained were either products that required special conditions during shipment (largely perishable foodstuffs) or low value and bulky commodities more appropriate to other shipping modes.

Subsequent experience has proven otherwise, and by some estimates container traffic worldwide was still growing by as much as 8.6 percent per year in 1989.[2] First, there were rapid advancements in refrigerator and climate-controlled containers in the late 1980s. By the end of the decade, all but a few meat shipments had been containerized, and containers were being offered that promised very reliable and precise humidity and temperature control (1–3 degrees centigrade). Furthermore, filters had been introduced to absorb damaging gases emitted within the container by perishables as they age. These technologies were reported to have extended the life of products such as mushrooms and apples within a container for more than a month and a half.[3] Similar successes were reported with the containerization and ocean shipment of such "difficult" cargoes as cocoa beans, coffee, rubber, tobacco, fresh flowers, and live eels.[4] Now that the problems of containerizing such cargoes largely have been solved, there seems little reason to perpetuate the break bulk-reefer ship system, with its high labor and handling costs. Thus it seems likely that many of the products currently carried in the holds of refrigerator ships will be transported in containers and that, as the reefer fleet continues to age, it will gradually be replaced, not by new vessels of any kind, but by refrigerated ocean containers carried on ordinary containerships, as well as by air freighters.

Second, many cargoes previously regarded as too clumsy or of low value for containerization are now being containerized. This is particularly likely to occur where load factors are low, usually backhauls on routes that have a rich one-way traffic flow. Thus, for example, unsawn logs are stuffed in containers in U.S. Pacific ports for shipment to Asian customers, and Nigerian cotton and ground nuts fill containers that might otherwise return empty to ports in North Europe. Plastic liners for containers have made it possible to transport a "dirty" bulk commodity and then quickly prepare the same container for a more conventional cargo. The containerization of low value break-bulk and "neobulk" cargoes is likely to grow in the 1990s, since the surplus of container carrying capacity will compel carriers to seek additional tonnage to help fill vessels and cover costs. However, traditional methods of shipping such cargoes are likely to remain dominant, since they usually will be less expensive, particularly when the volumes moved are substantial.

Third, the exports of newly industrialized and less developed countries will continue to be containerized rapidly. Many of these products have long been containerized in trade within the industrialized world, but they have had to be handled breakbulk in NICs and LDCs due to the absence of container handling equipment and adequate inland infrastructure. Recognizing the importance of containerization to participation in the mainstream global economy and, hence, to national economic development, many developing countries have made such investments high priorities.[5]

Although containerization of cargoes will continue, it probably will advance at a somewhat slower rate over the next decade than it has over the last two. Estimates of containerization are imprecise, but perhaps 60 percent of what had been break-bulk cargo earlier had been containerized between the later 1960s and 1985—an annual rate of around 4 percent. Even if one expects containerization to reach 85 or 90 percent of general cargoes by the turn of the century, the rate of expansion will necessarily be lower—in the area of 2 percent per annum. If this estimate is combined with expectations that world trade in general (including cargoes already containerized) will expand at rates not greatly different than those experienced during the past two decades (3–5 percent a year in terms of tonnage), it is reasonable to conclude that containerized cargoes will continue to grow, although less rapidly than in the past.[6]

Fleet Capacity, Supply, and Demand

Although containerized cargoes will continue to grow, the capacity of ocean container carriers seems likely to grow even faster, at least through the first half of the 1990s. Thereafter, the relationship between supply and demand is less clear, and outcomes are likely to be influenced by developments in the tanker and bulk sectors as well as industry and government actions relating to capacity, price, and service.

Through the 1980s, global container carrying capacity generally has run 20–30 percent ahead of demand. Furthermore, weak demand for new tankers and efforts by some governments to subsidize their faltering ship-building industries helped cause a decline in the price of new containerships in the late 1980s. This, in turn, stimulated a bubble of new building orders, particularly for jumbo-size vessels (2,000 teus and more) intended for the mainstream East-West trades between Western Europe, North America, and East Asia. By mid-1988, analyses of orders already on the books indicated that the carrying capacity in those trades alone would jump nearly 60 percent by 1991.[7]

As the decade drew to a close, ship lines such as APL, Evergreen, Hapag Lloyd, Nedlloyd, Maersk, Yang Ming, and P & O had either put into service or placed orders for fleets of faster (22–25 knots) gearless vessels in the 3,500–4,500 teu range. Many of these ships were designed so that they later could be "jumbo-ized" to the 5,000 teu range.[8]

Furthermore, expectations that older, smaller containerships would be scrapped as the larger vessels came into service have not been fulfilled. Instead, many of these "obsolescent" vessels have been purchased by other operators, often new entrants into the global container carrying market. Some of these ships remained in the mainstream services; others either became feeders to those services or were operated on other routes. In any event, many remained in the global fleet, exacerbating the capacity surplus.[9]

For all of these reasons, chronic excess capacity seems unavoidable through the early 1990s. What happens thereafter depends in part on what happens in the tanker and bulk trades. Although there were episodes in the late 1980s of strengthened demand for particular sizes or types of vessels in both trades, in general new building remained modest. Rather than build new ships, operators often sought to extend the useful life of existing tankers and bulkers vessels well into the next decade through extensive refits and long-term charter arrangements. A boom in either the tanker or bulk trades in the 1990s could alter the picture for containership construction by raising prices and causing orders for new vessels to be cancelled or postponed. Although experts currently see such a development as unlikely, various economic, environmental, political, or military events could cause a rapid change, particularly in the tanker market—as they have in the past.

Even without that, however, some analyses have suggested that the obsolescence of the tanker fleet is developing fast and that more than 300 tankers will have to be built as replacements during the 1990s. Should that occur, its effect, particularly in the second half of the decade, could be to drive containership prices up and slow the growth of capacity as the lines delay or cancel orders.[10]

Chronic excess capacity in container slots in the 1990s is likely to lead, as it did in the 1980s, to intensified competition and additional bankruptcies, mergers, and acquisitions among containership operators. Increased concentration of carrying capacity into a few giant ship lines is one clear prospect.

Other responses also may occur. Collective action by ship owners to limit capacity and price competition in specific trades or reach accommodations on market share are likely to be attempted. These may take either the traditional form of shipping conferences or involve new structures such as the Transpacific Discussion Agreement, which emerged in late 1988 as a forum for negotiating capacity limitations in the trade between Asia and North America. However, the likelihood of substantial excess carrying capacity worldwide means that agreements on mutual restraint—even among *all* the operators in a particular trade at a given moment—are unlikely to be very robust. Such agreements will be under constant threat from the possibility of new entrants in the trade and from the cost and profitability pressures upon each member to leave the agreement or quietly violate it. These pressures are likely to be reinforced by the continued skepticism of shippers and of regulatory bodies, in the United States and perhaps elsewhere, about the legality, equity, and efficiency of conferences and other competition-restraining initiatives by transportation providers.[11]

Governments could, of course, play a role in capacity restraint by encouraging their "national" or government-owned ship lines to operate profitably or die. They also could discourage excess shipbuilding capacity and create incentives that cause private vessel operators to rationalize their services and scrap rather than resell older vessels. The Japanese government's efforts in 1988 and 1989 to encourage Japanese liner companies to engage in mergers, routing realignments, slot-sharing, joint services, and capacity reductions is an example of one such initiative.

However, the record in this regard of European, American, and Asian governments has been erratic at best. Furthermore, the arguments in favor of maintaining (or even expanding) carrying and building capacity—nationalism, employment, and military security—remain politically potent. Therefore, it seems realistic to predict that the key governments are likely to continue to vacillate between policies that stimulate restraint and expansion—and in some cases to pursue both at once.

This leaves at least one other source of constructive response to the excess capacity problem, voluntary actions by individual carriers. The strategy adopted by Sealand, TFL, and Nedlloyd in restoring to service the former United States Lines Econships seems an example. The vessels' original design capacity of 4,482 teu was voluntarily reduced to about 3,400 teu of loaded containers. However, it is unclear how many other carriers will follow this example, since an important objective was to increase these vessel's operating speed from the 15–16-knot range (planned in an era of $30/barrel) to the 19–20-knot range so that they could compete more effectively with vessels being introduced by competitors in the North Atlantic that offered even faster passages.

Richard Gibney, the late editor of *Container Insight*, suggested another form of voluntarism—not an adjustment in capacity, but rather an adjustment in *attitude* about capacity. "A possible scenario, where every owner of every jumbo containership begins by looking for 99 percent slot utilization of his new craft from the maiden voyage onwards, really would be scary," he wrote. Gibney noted that the international airlines faced a analogous excess capacity problem with the rapid transition to jumbo jets in the 1970s. However, they responded by "generally work[ing] with load factors in the 55–75 percent range. . . ." He proposed that, rather than striving to fill every slot in every new jumbo containership from the outset, the ship lines adopt strategies that accept lower load factors initially, in anticipation of a gradual buildup from the continued expansion of world trade and from "their ability to lure more and more lower-rated big volume cargo into the box. . . ."[12]

Vessel Size and Speed

As noted in Chapter 5, it usually is more economical to make a ship larger than to make it faster. This is due to several relatively fixed relationships between hull shapes, construction materials costs, propulsive power requirements, and vessel speed.[13] Despite the fact that larger is usually more cost efficient, during the early 1980s many transportation experts suggested that there were logical upper bounds—geographic, commercial, and technical—on the size of containerships. One such bound was "Panamax," the breadth and draft limits of the Panama Canal, which imposed a carrying capacity limitation around 3,500 teus. The inference was that anything larger would be unable to transit quickly between oceans, and no shipowner would want that. This attitude quickly diminished as transPacific trade grew rapidly and landbridging across the United States increased, facilitated by the introduction of daily unit trains and double-stack railcars. Even a large ship line could make money operating in the Pacific only, and, if it wished to, it could provide faster round-the-world service than an all-water carrier by landbridging between the two oceans across the United States.[14]

With the Panamax limit hurdled, it appears that the geography of other major international waterways will permit containerships to grow considerably wider and deeper in the years to come. However, if vessels actually were laden to drafts of 45 feet or more, it would necessitate harbor channel and berth dredging projects, even in major container ports.[15]

One commercial concern was that there might not be enough cargo to fill much bigger ships (a point that has yet to be disproved, as noted above). Another was that containers would not be concentrated in sufficient volume at a small enough number of ports. However, as increasingly heavy concentrations of containerized cargo continue to be attracted to the largest load-centers, such as Hong Kong, Singapore, Rotterdam, and Los Angeles-Long Beach, that apprehension has subsided.

One technical concern was that the design capacity of yard equipment and the limitations of infrastructure in and around the container terminal created obstacles to the rapid loading and discharging of huge numbers of containers, thereby diminishing the advantage of very large vessels. However, by the end of the 1980s a few terminals (such as ECT Maasvlatke) had in place ship-to-shore cranes and yard equipment fast enough so that, if three of the cranes were ganged, a 5,000-teu vessel carrying 40-foot containers could be entirely stripped and fully reloaded in a day and a half. Furthermore, as mentioned in Chapter 4, computer and communications systems had been developed that could support the planning and implementation of such an undertaking at a high level of efficiency.

The designing of containerships with nearly double the current capacity also seemed technically questionable in the mid-1980s. One problem was that such vessels would be wider than the outreach of existing ship-to-shore cranes. However, by the end of the 1980s, many terminals had purchased cranes that could work across 16 rows of containers on the ship, accommodating the largest vessels being planned. Another concern was that these very large vessels might lack structural strength athwartship and that failures might result. By the end of the decade, some adjustments in design as well as experience with the first jumbos delivered had provided reassurance in this regard.

As the 1990s begin, therefore, most of the concerns that previously had constrained containership size had been satisfactorily addressed. Thus it seemed likely that the decade would witness vessels of 5,000 teus and more, ships twice as large as the largest being built just 10 years before.

Furthermore, it appears that the vessels built, at least during the early part of the 1990s, will be somewhat faster than those predecessors. Trade-offs between speed, fuel consumption, and vessel productivity are hard to project in a world of volatile fuel prices, and during the 1970s and 1980s the price of a barrel of oil first quadrupled in a year, then doubled in a year, then dropped nearly 65 percent, and, finally, increased more than 75 percent. Nevertheless, it appeared that ship operators were opting for vessels 2–5 knots faster than the previous generation in their new ship ordering at the end of the Eighties.

Vessel Types and Characteristics

It seems clear that the trend toward nonself-sustaining or gearless cellular containerships that was obvious in the 1980s will continue into the next decade. With few exceptions, all of the jumbo vessels ordered at the end of the decade were of this type, and they seem certain to dominate the major East-West trade routes in the years to come. It is probable that vessels of this size and type will comprise the bulk of the world containership fleet in tonnage terms by the early 1990s.

However, container vessels with their own gear as well as RO-ROs and LO-ROs are unlikely to disappear from the world fleet in the foreseeable future. It is likely that many feeder ports and ports in developing countries will have no large ship-to-shore cranes, or only one or two that are older and slower. Thus the medium-size and small vessels calling at such ports will rely, either totally or partly, on their own loading and unloading capabilities.

Second, RO-RO vessels and others with their own discharging capabilities are likely to remain attractive for military purposes (as well as certain civilian uses, such as the shipment of heavy construction equipment). Therefore, some governments are likely to subsidize or otherwise encourage their construction and operation by private firms or state-owned ship lines. In contrast, conventional containerships have more limited military value, since they must be unloaded at a wharf with a container crane or by other vessels or barges equipped with heavy-lift cranes.

Furthermore, RO-RO and LO-RO vessels are likely to be well suited for certain short sea operations, for example, in Western Europe and parts of East Asia and the Pacific. Such operations, in fact, will be container ferries and, as such, will prefer the high speed with which containers already on chassis can be debarked and depart the terminal upon arrival.

The next decade is also likely to see increased use of barging for shortsea and coastal feeder services as well as on inland waterways. In situations where ports are equipped with adequate ship-to-shore cranes, the use of seagoing barges and tugs or pushtows may well be preferred to self-sustaining containerships, whose capital and operating costs are usually much greater. This pattern already is well developed along the U.S. East Coast and between the U.S. and the Caribbean. It seems an economically attractive alternative for feedering elsewhere.

If the industry were orderly and entirely rational, one might predict the emergence of a pattern in which three different types of vessels provided three distinct but interconnected types of service: (1) fast, huge, gearless transoceanic ships that call at only one or two load centers on a range, loading and discharging a thousand containers or more at each call, (2) RO-RO and LO-RO vessels (as well as older and smaller containerships) that served secondary ports, lower volume trades, and LDCs; and (3) pushbarges, tows, and smaller combination ships that carried 50–500 containers at a time and provided coastal, shortsea, and feeder services between load centers and "outports." Such orderliness is unlikely, however, if a large excess capacity globally persists and if the industry and governments are unable to manage it effectively. In that case, all sorts of "crossovers" of vessel types from one kind of service to another are likely to persist, as operators search for routes that will increase load factors and generate some profit for the owners.

On and Over the Horizon

As the twenty-first century begins, more radical approaches to the ocean carriage of containerized cargo may appear. Ideas about how to provide faster or more efficient service (some of which have lain dormant for years in the minds and file drawers of designers) may emerge and become realities.[16]

Although conventional containership hulls have the potential for speeds perhaps 40 percent faster than the jumbo vessels being built today, the commercial feasibility of such speeds is dependent on the development of propulsion systems that are economical, compact, and much more powerful. At present, only nuclear technology offers the prospect of such a breakthrough. Various accidents and altered perceptions about the long-term cost efficiency of nuclear power have diminished its attractiveness. However, if a vessel could operate at 40 knots and cross the Atlantic in three days, it would help fill the great gap in speed of delivery that now exists between ocean carriers on the one hand and air freight on the other.

A second possibility for increasing speed is a vessel that moves in one medium only.

The maximum speed of a conventional (displacement) vessel is determined primarily by the wave resistance that its hull encounters at the interface between air and water. Thus a vessel able to travel either entirely below the water surface or entirely above it, can attain much higher speeds. Experiments with submersible tankers are foreseeable, for example, for use in ice-bound areas. However, container-carrying submarines are less likely, partly because of the high cost of building hulls that can resist underwater pressures and the large powerplant and high fuel consumption required to drive a vessel large enough and fast enough to make the investment worthwhile.

On the other hand, experiments with containerships that operate *above* the water surface could occur before the end of the next decade. In fact, "airborne" passenger-and-car ferries and sightseeing vessels of several hundred tons are already in regular service all over the world. These craft are of two different types—hydrofoils and hovercraft (surface-effect vessels) that ride on air cushions.

The application of either technology to cargo vessels is attractive because hydrofoils are theoretically capable of 60 to 100 knots and air cushion craft, of speeds even higher. Thus either technology, if priced appropriately, would broaden the choices available to transportation consumers. Such vessels could move cargo two to four times faster than the fastest displacement ships but could accommodate cargo that was heavier or bulkier than generally appropriate for air freighters. Surface-effect vessels would have the additional advantage of requiring no berth space, since they could move up a gradually inclining shoreline or ramp directly to the terminal exit gate or some other inland location. Problems such as propulsion, fuel consumption, and heavy weather stability mean that *transoceanic* hydrofoils or hovercraft are unlikely in the short term. However, the introduction of smaller versions capable of carrying 50 to 100 high priority boxes on ferry runs and short sea routes—for example, in Northern Europe, the Mediterranean, and East Asia—seems a realistic expectation.

Innovations in displacement vessel designs are also foreseeable. One possibility is an advanced RO-RO design with very high loading and discharging rates. Such a ship might have multiple openings and ramps, rather than a single one as is now the case. However, it would require substantial investment by terminals in specially constructed berths. These would have to be configured like a small boat slip or drydock, with an opening at one end and dock area on the other three sides that would accommodate the multiple ramps and all their traffic flows.[17]

Another possibility is the introduction of "megacontainers," which would take the basic concept of containerization—consolidating many small parcels of cargo into a single, standardized parcel—to another level. If faster port calls remain desirable, it seems logical to rethink containership design in terms of leaving and picking up whole chunks of the vessel—each chunk holding dozens of containers. The departing megacontainer could be completely prestowed before the vessel's arrival, and the arriving one could be "stripped" after the ship departed.

The technology for moving these megacontainers off and on the ship already exists elsewhere in the maritime industries. For example, shipyards utilize gantry cranes with capacities up to 1,000 tons, and "Seabee" and LASH barge carriers routinely float or lift barges of 500 to 1,000 tons on board.

However, the practicality of megacontainers would be determined in part by whether they made discharging and loading a vessel substantially faster than ganging several of the high-speed cranes that already exist.[18] Widespread acceptance also would depend on the terminals' ability to efficiently and rapidly stuff and strip these megacontainers, since the ultimate customers—shippers and consignees—care about speed from door-to-door, not port-to-port. Furthermore, like containers and chassis today, the repositioning of megacontainers would present a challenge, especially since the cost of moving empty ones would be high.

Other, even more radical designs exist. These include huge container-carrying catamarans ("trisecs" or "SWATHs" in the jargon of their designers) and containerships that would be "made up" for each voyage out of separate components. The latter might consist of a stern component that would carry the propulsion machinery and the crew's working and living quarters, a bow, and other pieces that would hold the containers. Although intriguing, each of these ideas still confronts difficult engineering obstacles. Furthermore, most seem to require even heavier investment by terminals—in the design and acquisition of entirely new cranes and other equipment, as well as major construction projects to reconfigure berths to accommodate the new vessel designs. Thus they seem unlikely to be built until well into the next century, if ever.[19]

Crews of the Future

During the 1990s it will be possible to operate an ocean-going containership with no one onboard. Instead, all relevant data about the vessel's exact location, the status of its own systems, and the conditions in which it was operating (such as weather, sea state, and traffic) would be transmitted automatically to land-based personnel thousands of miles away. These personnel—expert engineers, technicians, and navigators—would monitor their computer terminals, printouts, and TV screens, make all operational decisions, and transmit commands back to the vessel that cause the necessary adjustments. Multiple backup systems would insure against mechanical failure.

Although reducing crew sizes and labor costs are attractive to the carriers, concerns about safety and environmental protection (heightened by such incidents as the grounding of the *Exxon Valdez* in March 1989) make it unlikely that unmanned containerships will become a reality anytime soon. Even so, the most modern containerships of the 1990s will have crews less than half the size of their predecessors. By 1989 Transatlantic Shipping Company (Sweden) was operating a 2,900-TEU vessel with a crew of 12, Hapag-Lloyd also had announced plans to operate with crews of 12 within three years, and ACL was reportedly experimenting with crews as small as nine. Furthermore, there were already several vessels in service (including two Hapag-Lloyd jumbo containerships) that could be "singlehanded"; all their systems could be monitored and controlled by one individual on the bridge. Just how far crew reduction will go is uncertain, but national and international maritime agencies may well adhere to the view of one insurance underwriter who declared, "Ships crewed by only four or five people would be only a computer chip away from disaster."[20]

However, it seems clear that containership crews will become even smaller than they are now, and the nature of the life and work will continue to change. Some experts expect the traditional hierarchy onboard to disappear as containership crews small teams of perhaps six to 10 highly qualified technicians who are cross-tra the tasks of both deck officers and engineering officers. Although each crew member responsibilities will increase, they will entail primarily scanning and monitoring instruments and screens for signs of abnormalities, rectifying them, and performing complex but standardized maintenance procedures. Furthermore, the vessels these crews will sail are likely to follow even more fixed routes and make even shorter port calls than currently, and life onboard could become even more isolated and routine. To offset the adverse effects and to fill the increased periods of freetime, all sorts of leisure time and personal and professional activities will be necessary. Wives sometimes will sail with their husbands (as is already the case with a few lines). In addition, many vessels may well be double-crewed (as is done now with some naval vessels). Fresh crews will be rotated every three or four weeks, and some of the time ashore will be spent in advanced training and in performing administrative and technical tasks for the vessel or the company as a whole.

Labor costs can be lowered not only by using smaller crews, but also by using lower cost personnel. In principle the containership fleets of the industrialized countries have been prohibited from doing this by laws or labor-management agreements that preserve all jobs on a vessel to home country nationals. In practice some carriers have circumvented these provisions by using foreign registries (flags of convenience), which impose no requirements regarding the nationality of their crews. Currently, the use of crews of mixed nationality is being considered by a number of Japanese-flag carriers as a means of lowering labor costs. Such initiatives are likely to be considered elsewhere in the coming years, along with smaller crews.

TERMINALS AND PORTS

Stack vs. Chassis

Chapters 2 and 4 examine the trade-offs between operating a chassis-oriented terminal and a stack-oriented one. One inference that might be drawn from those analyses is that increasing customer demands for speed and reliability may militate in favor of chassis-oriented operations in the future, particularly at terminals where land is readily available and labor costs associated with moving boxes in and out of stacks are high. On the other hand, however, a number of factors suggest a trend toward stack operations, especially among larger marine terminals.

First, an increasingly large proportion of the export containers received by those terminals will arrive without chassis. This is because it is likely that the proportion of containers arriving and departing by unit train, barge, and feeder vessel will rise. (Although the number of containers coming and going by truck will remain substantial, the proportion will decline.) Similarly, large terminals are more likely to attract the port calls of the 4,000–5,000-teu cellular containerships, and all the containers they discharge (unlike RO-ROs and LO-ROs) will be without chassis.

Ocean Container Transportation

ship line chassis are a rarity. Containers that arrive
must be taken off them promptly anyway.
easingly face space constraints as well as rising ac-
. Using the land you have more intensively, rather than
chassis-oriented operation, will seem logical to many

explains, better computer programs for planning and per-
ve yard activities will become available. These will make
rations more acceptable for terminal operators, carrier and
shippe

Improving Terminal Efficiency

Beyond this, the next decade is likely to see increased experimentation with three different approaches to improved container terminal efficiency and faster customer service. The first requires no changes in technology but instead involves operating the terminal seven days a week and 24 hours a day. From the customer's viewpoint this means that containers can be dropped off and picked up at any time. From the terminal operator's perspective, around-the-clock operation is a way to increase terminal capacity where physical expansion is impossible or very expensive. To be cost effective, however, this innovation often will have to be accompanied by modification of traditional labor contract provisions about overtime pay and working hours.

The second innovation involves applying technology that permits the handling of more than one container at a time, and the third would employ more integrated and automated systems to move *individual* boxes between the storage area and the vessel.

Although it is rarely done with loaded containers, some existing ship-to-shore cranes, transtainers, and straddle carriers are capable of "twin-20" lifts, in which two 20-foot containers stacked on top of each other and locked together with corner pins are moved as a unit. There seems to be no theoretical reason why larger, heavier (and more expensive) terminal equipment could not lift blocks of four or more containers simultaneously. Other systems for multiple container movement exist, for example, ECT's container "trains," which pull five 40-footers at once using heavy-duty tractors and the LUF-system in which a large tractor tows a platform carrying blocks of four or more containers. It is foreseeable that some of these technologies could be utilized in the loading and unloading of the ocean-going "megacontainers" mentioned earlier.

Although a number of integrated ship-to-storage-stack systems for handling containers have been designed, to our knowledge only one, used by Matson Lines, is in regular operation.[21] These designs involve an uninterrupted linkage between the ship-to-shore crane and storage areas. That linkage may be an oversized transtainer that receives the container directly from the crane, a conveyer belt, an overhead monorail system, or a mechanical "merry-go-round" that flows to and from the vessel continuous and, at least in theory, might be programmed to simultaneously discharge and reload the same hatch on the vessel.

There has been understandable reluctance in the industry about embracing these integrated systems. First, the investment costs are enormous, and they necessitate a permanent and very radical alteration in terminal operation. Furthermore, the inflexibility inherent in such systems raises questions about their operational and commercial practicality. What happens, for example, when there is a problem with one box? Or when the equipment (or the supporting computers) suffers a breakdown? Can such systems serve the needs of multiple ship lines that operate vessels of somewhat different designs, make different sequences of port calls, have their own operating preferences, and even use containers of different sizes? Will the production runs of containers going on and off even very large load-centering vessels be big enough to justify use of such equipment? And, finally, will such systems be any faster than working such vessels with two or three dual trolley cranes supported by appropriate yard equipment and management systems?

Skepticism about the answers to these and other questions suggests that integrated systems will continue to be quite rare. As in the Matson case, they are most likely to be installed at dedicated terminals run by ship lines whose routes are very stable and involve only a few high volume calls at either end of the ocean crossing.

A different approach that uses advanced computer and automated guidance technologies to move individual containers at very high rates of speed was being tested by ECT in 1989. The system was designed for use in serving one line, Sealand, which will be load-centering at ECT-Maasvlatke and providing feeder service from there to other ports. The system utilizes existing dual-trolley high-speed wharfside cranes. However, these cranes are served by a fleet of "automated guided vehicles" (AGV)—essentially, unmanned straddle carriers that run to and from the stacks. At the stacks, the containers they carry are moved and stored by oversize yard gantry cranes that also are unmanned. Preliminary tests have shown that, under ideal conditions, the system can run at rates up to three times those of convention terminals while avoiding some of the rigidities inherent in systems involving more tightly integrated equipment. Its greater flexibility suggests that it might even be applied to some multiple carrier situations in future years.[22]

Two other innovations that would enhance terminal efficiency and *are* likely to become widespread in the next decade lie in the areas of cargo inspection and computerization. Tests are currently being conducted on passive methods for examining the chemical composition of the contents of a container without stripping it. Such methods offer the prospect of both enhancing the detection of illegal narcotics or terrorist bombs within containers and of reducing the frequency of devanning, with the expense and delay it causes the shipper. However, the identification of other contraband—for example, conventional weapons, counterfeit goods, and restricted technology—will, it appears, continue to require manual searches.

Current and future advances in computer technology also will have an impact on terminal management. Microprocessors and other computer hardware have grown rapidly smaller and more powerful. Desktop computers can now perform functions that only 10 years ago required a large mainframe—and at a fraction of the cost. Thus smaller terminals will be able to take advantge of sophisticated techniques for tracking and organizing containers, and larger terminals will be able to distribute computing power to users such as yard supervisors, instead of concentrating it in centralized data processing centers.

Computer software, especially for microcomputers, lags behind hardware developments, but it is also becoming more sophisticated. New developments in artificial intelligence and expert systems mean that programs can be created to assist in stack layout, stowage planning, and work scheduling. Heuristics and optimizing algorithms can replace guesswork and help eliminate inefficiencies, for example, in yard management and stowage planning. Again, the cost of these systems is low. Thus even small terminals will be able to take advantage of this technology, and larger terminals will be able to make such techniques available to the lowest level decision makers.

Finally, improvements in both hardware and software are making the technology accessible even to relatively inexperienced users. For example, sophisticated menu structures and decision tree techniques can lead the user through complex computer applications, even though the user has a minimal knowledge of computers. This means that within a few years the computer will be as easy for most poeple to use as the telephone. As this computing power becomes available to more and more users on container terminals, many of the things discussed in Chapter 4, such as capacity utilization and scheduling, will be done more efficiently and at less cost than today.

Terminal Labor in the Future

In the future even fewer longshoremen will be needed to more even greater volumes of cargo, continuing the trend of productivity improvement that began with the advent of containerization. Furthermore, few of the workers that remain (perhaps only lashers, stuffers, and strippers, and some maintenance personnel) will do the physically demanding tasks that have typified dock work in the past. Instead, container terminal workers will be responsible for operating and maintaining the increasingly sophisticated and expensive container handling equipment and the communications apparatus and computers that comprise the central nervous system of the modern container terminal.

This means dramatic increases in responsibility for some. In the past, dropping a crate of breakbulk cargo might cause $1,000 worth of damage. In the future, dropping a loaded container might destroy $100,000 worth of cargo, endanger a $5 million crane, and cripple a key part of the whole terminal's operation. Similarly, in the closely integrated, high-speed world of intermodal transport, mistaken data entries or computer breakdowns could cause expensive disruptions and damage relations with important customers. Furthermore, as the prior section has suggested, longshoremen will not only do operational work, but some of them will also make decisions about when and how that work is done, utilizing the capabilities available through the computer.

These tasks will be performed in the context of continued pressures to cut costs and increase productivity that will result from continued (and sometimes intensified) competition between carriers and between ports as well as from the requirements of increasingly tight intermodal linkages. Although containerization has forced terminals to become increasingly capital- and technology-intensive, labor costs continue to comprise more than half the operating costs at a major container terminal.[23]

In such an environment all sorts of changes in traditional practices seem essential. The workers themselves must be more highly educated and must undertake more extensive training, both before they go to work and periodically thereafter. Changes also seem necessary in the arm's length relationships between terminal managers and unionized workers; in rigid job jurisdictions and work rules; in the irregular and uncertain nature of terminal employment; in expensive, often self-defeating job preservation schemes; and in the resistance of both labor and management to consider new ideas and new technologies. The ports and terminals most likely to be successful in the future are those that are able to achieve these transitions quickly and smoothly.

Ports and Port Authorities

Chapter 1 painted a picture of intensified competition among ports during the 1980s, propelled in part by government deregulation and privatization and in part by the pressures imposed by ocean carriers who themselves faced rising costs, excess capacity, and unsatisfactory profitability. That picture will persist in the 1990s, with some modifications.

The decade is likely to see some "shake out" in the competition among ports for load-centering status in the mainstream routes, particularly in the industrialized West. As the costs of continuing to compete keep rising (in terms of new equipment, construction, and dredging), some unsuccessful competitors are likely to bow to fiscal realities or public pressures, curtail new investment, and, in effect, abandon the race. Some of these may well develop specialized niches in secondary trades or in specialized cargoes. Others may accept feeder port status, taking comfort in the fact that the employment and personal income effects per ton are far higher for stuffing and stripping, consolidating and handling breakbulk than they are for simply handling containers.[24] Some, burdened with expensive, underutilized ship-to-shore cranes and other equipment, will sell them off to the "winners" or to newly emerging ports in developing countries. Other ports will regard these and other terminal investments as sunk costs and offer their use at low prices as a means of attracting customers. These customers might be smaller carriers, or they might be large ones who have become convinced there are diminishing returns to calling at the same regional load center as everyone else and are willing to try an adjoining port that is not saturated with competitors.

The traditional considerations noted in Chapter 1 are likely to continue to be significant. As vessels get even bigger and more expensive, the operators' desire for even faster circuits and higher annual utilization rates should reinforce the advantages of ports with speedy access to major shipping lanes and channels with depths of 45 feet or more. Similarly, ports with populous, affluent industrialized cargo bases of their own (as well as good intermodal connections deep into their continent) will be best situated to emerge as principal load centers. In an era of 5,000-teu vessels, calls at only one or at most two ports on a range are increasingly likely, with feeder, landbridge, and minilandbridge connections to other locations.[25]

The bubble of investment in terminal and port development during the 1980s will continue in the coming decade. New investment will have to be widespread, of course, if some of the changes in vessel design or container-handling technology described earlier become realities. Otherwise, there is likely to be a proportional shift in such investments toward the developing countries, particularly those of Asia and the Pacific. According to some forecasts, by the year 2001, six of the world's top 10 container ports will be in the newly industrialized countries of East Asia.[26]

However, even within the same region, investment patterns are likely to vary with the role of government and the perceived purpose and value of the facility. Other factors being equal, investments are likely to be larger and more sustained (and sometimes less rational) in places where decision makers measure the value of a port or terminal not strictly in terms of profitability or return on investment, but rather as an investment in basic infrastructure important to economic development, job creation, and the attraction of new investment.

The port authorities' domain, which generally broadened during the 1980s, is likely to stabilize and, in some instances, narrow in the years to come. Indeed, that pattern already seems visible. Some of the activities in which port authorities involved themselves during the 1980s were undertaken because no one else was performing them. In some instances, these activities (for example, the operation of an export trading company, a chassis or container pool, or a container freight station) have been privatized, spun off as separate enterprises, or reassigned to more appropriate government agencies. Other activities thought to be helpful in attracting new business to the port have been shown by experience or evaluation to be ineffectual and have been curtailed or abandoned.

Furthermore, in a number of developed and developing countries where government-run terminals have been shielded from competition and are particularly inefficient, privatization (through long-term contract or outright sale) is under consideration.[27] Port authorities will have particular responsibilities in such places and in others where there is no meaningful competition among container terminals (for example, a small country that has only one terminal that handles all its containerized cargo). These responsibilities will include measuring and evaluating terminal efficiency and productivity and devising mechanisms to stimulate them in the absence of competitive pressures. Port authorities that fail to do so, in effect, will be raising costs for exporters and consumers alike, and thereby slowing national economic development.[28]

Finally, it appears that the ports are likely to be under continuing pressures from "gentrification" and from the environmental concerns of their neighbors. Ports, particularly those that are within or contiguous to large cities, will continue to find waterfront property in demand for residential, commercial, and office use. This will limit terminal expansion or make land acquisitions very expensive. Although a few ports will respond to this concern by preemptive purchasing of land well in advance of expansion projects (like Los Angeles-Long Beach) or will obtain governmental protection for potential future terminal sites, most will not be able to do so. In addition, as described in Chapter 1, many terminals are under pressure about congestion and road damage, dredge spoil dumping, dangerous cargoes, air quality, noise, and other environmental considerations. These pressures are unlikely to abate in the years to come. In fact, they probably will intensify, not only in the developed countries, but also in the developing ones.[29]

Environmental concerns and gentrification will force ports to develop terminal sites on the periphery of major cities. Simultaneously, the ports' primary customers, the ship lines, are likely to have an increasingly strong preference for sites on or near the open sea. Together these forces are likely to lead to the development of new terminal sites, often on newly reclaimed land near the coast line (as is the case with ECT's Maasvlatke terminal). An alternative will be the construction of container terminals on artificial islands—within an existing harbor (as in Tokyo and Kobe) or even off-shore. These islands will be connected by dedicated causeways, tunnels, and bridges to rail and highway systems on the mainland.

INTERMODALISM

Intermodalism Accelerates—The Air Link

Although there is some controversy about the prospects for further development of intermodalism,[30] we expect intermodalism to accelerate in the 1990s—barring major economic, political-military, or environmental disasters. The emergence of an entirely new mode of cargo movement during the decade seems improbable. However, it is likely that existing technologies will be modified or applied in ways that fill out the spectrum of possibilities available to the consumer of international transportation services, in terms of speed and cost. An important part of this process, already underway, is the competition between modes. One imminent aspect of this competition is the shift of higher value containerized cargoes from ships to planes. This shift is occuring for several reasons. First, the proliferation of FAX machines and other means of electronic transmission has damaged prospects for the overnight letter carriers and led them to pursue the market for parcel transport. This, in turn, has created a capacity surplus in the air freight industry, touching off the pursuit of lower valued cargo—cargo that normally moves in ocean containers.

Second, many containerized cargoes are becoming both more valuable and lighter. These cargoes include more expensive consumer goods as well as the components and products being moved to and from assembly plants in developing countries. As freight becomes lighter and more valuable, it becomes a more likely candidate for air carriage.

Third, as mentioned earlier, the customer cares about *door-to-door* speed. No matter how fast the ship, if its cargo encounters delays in the port of arrival, if inland transportation systems are underdeveloped or unreliable, or if the shipper and consignee are both a thousand miles from the nearest seaport, air freight may, in fact, be cost efficient.

Furthermore, in developing parts of the world perhaps more than elsewhere, difficult topography and inadequate ground infrastructure may well make ocean-air intermodalism an attractive alternative. If so, efforts to make air and ocean containers compatible or interchangeable, abandoned years ago, may be resumed and pressed to a successful conclusion.[31]

Intermodalism Accelerates—The Rail Link

Chapter 1 describes the appearance of double-stack and unit trains as well as the efforts of some operators in the United States to develop domestic containerized cargoes as a means of minimizing empty unit train back hauls. The continued development of "domestic" containerization seems a necessary corollary to the rapid expansion of transoceanic containerization intermodalism. In fact, the term "domestic" will be technically inaccurate, since some of these containers will cross national borders (for example, within the E. C. or between the United States and Mexico). However, they will not have a water leg.

The pace of this development is important to the global expansion of containerization because the capital investment and operating costs of intermodal systems are so heavy that they are hard for railroads and other operators to justify on the basis of transoceanic cargoes alone. Although ocean containers may never fully replace over-the-road trailers, domestic or pure overland moves are likely to become an important way for container owners and transporters not only to smooth imbalances in international flows, but also to spread costs and maximize net revenues.

As part of this process, railroads are likely to encourage their customers to switch from "piggyback" shipment of trailers on flat cars to containers carried by double-stack trains. Similarly, the next few years are likely to see the adaptation of the "road-railer" technology mentioned earlier to ocean containers. That technology is currently in use for trailer chassis only. The adaptation would involve building container chassis that had two sets of wheels—one steel, the other rubber. This would allow a container to be converted, in effect, from a rail car to an over-the-road truck body in a matter of minutes.[32]

New Routes and Methods

In addition to the innovations described earlier in this chapter, other intermodal prospects for the next decade include:

- Expanded use of rail ferries carrying whole unit trains on shortsea routes.
- Emergence of new major landbridge routes. These include the Northern Europe-Mediterranean landbridge already developing, the underutilized trans-Siberian route between Europe and the Pacific (if *perestroika* leads to substantial improvements in the operation of the Trans-Siberian Railway,[33] and a landbridge across Central America (if political difficulties interrupt movement through the Panama Canal).
- Increasing use of bridge-tunnel solutions as alternatives to ferry service on short, high volume crossings. Most notably, the "Chunnel" between France and the United Kingdom is projected to go into service in 1993. Other, more remote candidates for such links include the EC and Scandanavia (across the Denmark Strait), Western Europe and North Africa (via Gibraltar), and the Middle East and Europe (via the Bosphorus or Dardanelles).
- Experimentation with the much shorter Arctic routes between parts of East Asia and Europe or the East Coast of North America.[34]
- Direct rail-ship transfers of containers, especially at terminals where one ship line has a dedicated, regular unit train operating to and from a single, high cargo volume inland destination.

Technology and Standardization

We expect many of the intermodal developments we have described to proceed rapidly in part because technological systems will be in place to accommodate and accelerate them. This expectation therefore implies the continued adoption by key government agencies (in customs, taxation, and trade and transport regulation and promotion) of advanced computerization and communication capabilities; the continuation of current progress toward international EDI standardization and harmonization; the widespread adaptation of the new technology for cargo and equipment identification and location described in Chapter 4; and the expansion of existing trends toward container leasing and chassis pooling.[35]

The issue of container equipment standardization remains a major uncertainty. Some experts argue that the current norm in ocean transport—the standard 20- or 40-footer is an optimal compromise between the ocean carrier, for whom it is almost too small, and the shipper and inland carrier, for whom it sometimes seems too big. However, many ocean carriers and box leasors continue to offer nonstandard heights, and several major carriers, such as APL, Sealand, and Maersk, offer larger than 40-foot lengths. Especially in the United States and Western Europe, transportation firms interested in attracting "domestic" as well as transoceanic cargoes are interested in larger cubes that can compete with 48-foot over-the-road trailers. However, in situations where air-sea intermodalism seems a realistic prospect, even the 40-footer is likely to be too large and heavy—as well as the wrong shape—for cargo planes. Thus whatever standard may be adopted by the ISO, it seems likely that differences in carrier and customer preference mean that substantial deviation will continue.[36]

The failure to standardize more completely has important implications for some of the potential innovations we discuss here, since it almost certainly will constrain interchangeability and flexibility of operation. Will the same vessel—or the same "megacontainer"—be able to accommodate containers of five different lengths (say 20, 35, 40, 45, and 48 feet) or three different widths, without time-consuming adjustments and wasted space? Won't the utilization of other multiple-move methods—LUF frames, multiple container lifting cranes, etc.—be impeded or even defeated by such variety? Will integrated ship-to-storage systems, such as Matson's, be able to function in the face of such diversity? Will equipment variety complicate various forms of rationalization—pooling, slot charting, joint services, mergers, and acquisitions—which are likely to be necessary, if the industry is to respond efficiently to future changes in market conditions? The answers to these and other questions suggest that the pace of innovation and change in the industry may, in fact, be correlated significantly with equipment standardization.

Global Commercial Realities

The driving forces behind the continued development of intermodalism are global commercial realities. In the 1980s, JIT inventory management techniques and "quick response" production adjustment and shipping schedules were innovations. In the 1990s they will be nearly universal. Product life cycles will continue to shorten. As they do, minimizing inventories of potentially obsolete products will become even more important. Consumers in an ever-widening number of countries increasingly will demand current fashion, product quality, and producer responsiveness. Components and raw materials will be sourced and assembled at different locations around the world, and the resulting end-products will be sold to increasingly affluent populations all over the globe.

In such a world, the most successful transportation firms are likely to be those that profitably can provide a global distribution capability. They must offer not just cargo carriage but a total logistics package that provides frequent and reliable service for goods moving between any origin and destination.

Some of the earliest firms to move in this direction, as noted in Chapter 1, were American, namely CSX-Sealand and APL. They will be joined and perhaps overtaken in the 1990s by others who can respond effectively to the market's requirement for door-to-door management of shipments and for a continuum of types of service in terms of speed, frequency, dependability, and cost. Most will operate their own equipment across multiple modes but will also purchase services or enter into joint ventures in transportation modes or geographic regions where others are better established. The aspirants include a number of European firms that can be expected to use as their base an increasingly integrated European economy. They are Merzario/Mantovani, Maersk/Moller, Nedlloyd/Van Gend, and Loos, Bilspedition/Transatlantic and Schenkers/German State Railways. Other possibilities include several Asian firms that already blend air and ocean transport interests, namely Hanjin/KAL, NYK/K Lines/Nippon Cargo Airlines, and Evergreen, as well as the American air freight giants, UPS, and Federal Express/Flying Tigers.[37]

Strategic Management

By the beginning of the 1990s, the carrier, the terminal operator, the port authority, and the intermodal operator all lived in a vastly more complex and dynamic environment than they did 10 years before. Not only did that environment include many new forces, but it also imposed all sorts of new interconnections and interdependencies. Executives in each organization confronted the challenge of maintaining efficient, profitable current service, while planning for a continuously turbulent future.

Where complexity and change are great, uncertainty and risk are great. And in this industry, with its heavy capital costs, the risks have very large financial, organizational, and personal implications. Intensified competition has been translated in part into a contest to improve facilities and equipment. Those improvements often have long lead times, and, thus, carriers, ports, and terminal operators are forced to take major investment decisions well in advance of the reality for which those investments are designed. Some may be mistaken or mistimed, risking the health or even survival of the entire enterprise. Yet to stand pat may well be just as risky. Understandably, therefore, the industry is seeing a new emphasis on strategic management. Carriers, terminal operators, ports, and intermodal conglomerates have moved increasingly to the use of all sorts of forecasting, planning, and decision-making tools—data bases and systems for monitoring the various components of their environments, econometric models, computer-based simulations, and sophisticated master planning procedures. Given their situation they have no choice but to apply these tools for analyzing, forecasting, and designing their futures.

However, in an environment where factors are so interconnected and turbulence is so great, even the most intense analysis, the most thoughtful planning, and the most closely reasoned decision making may not assure success or even survival. Those who survive are also likely to be the possessors of size and financial depth, flexibility, auspicious timing, and good fortune.

SUMMARY

This chapter explores the future of ocean container transportation through the 1990s and beyond. Containerization of cargoes is predicted to continue to expand, although at a slower pace than in the past. However, the capacity of ocean container carriers seems likely to grow even faster, at least through the first half of the 1990s, and orderly responses by the industry and by governments to the imbalance between supply and demand appear unlikely.

The jumbo vessels entering the container fleet in the next few years are likely to reinforce the current trend toward much larger, somewhat faster ships that are nonself-sustaining. However, important niches will exist for RO-RO and LO-RO vessels and for barges. Looking further ahead, container-carrying hydrofoils and surface effect vessels may provide high priority service on ferry runs and short sea routes, and megacontainers holding dozens of standardized containers may appear at the largest load centers.

Many container terminals are likely to adopt techniques that permit the movement of multiple containers at once or the more rapid movement of individual containers, and new cargo inspection techniques and computer-based decision support systems also will enhance terminal efficiency. The nature of work to be done at the terminal will continue to change, necessitating a smaller and more highly trained work force and an end to traditional rules, practices, and attitudes on the part of management and labor alike.

In some regions, the competition between ports to become load centers will end, but many of the "unsuccessful" will pursue alternative strategies successfully, finding other profitable niches. The domain of port authorities generally will stabilize and, in some instances, narrow as some activities undertaken in the 1980s are spun off to other agencies, privatized, or abandoned as ineffectual. Ports in developed and developing countries alike will find themselves under increasing pressures over environmental concerns and from the forces of gentrification.

Intermodalism will accelerate, and the competition for international cargo between air and ocean carriers will intensify. Containerization of "domestic" cargo will increase, and new landbridges and bridge-tunnel routes may emerge, but problems of equipment standardization may well impede the application of advanced cargo movement technologies. The most successful firms are likely to be those that develop a truly global distribution capability and learn how to profit while providing a continuum of services in terms of speed, quality, and cost.

Notes to Chapter 7

1. W. B. "Bruce" Seaton of American President Companies, as quoted in *Container Insight*, Report No. 18, November 1988.

2. J. Smagghe, general manager of Le Havre Port Authority, at a meeting of the International Association of Ports and Harbors (IAPH), Miami Beach, as quoted in *The Journal of Commerce*, May 1, 1989.

3. *Containerisation International* 23, no. 3 (1989): 22.

4. For more on cargoes requiring refrigeration, see *The Reefer Market: Trends and Prospects in Refrigerated Cargo Trade and Shipping, 1985–95* (London: Drewry Shipping Consultants, 1988); and *Refrigerated Transportation* (London: Container Marketing, 1988).

5. For example, see J. Jacobs, "Far East Ports Adopt Bold New Strategies," *The Journal of Commerce*, April 28, 1989. Marine container terminals and related infrastructure improvements also have received high priority from the World Bank and other international organizations for development assistance.

6. Organization for Economic Cooperation and Development (OECD), *Maritime Transport: 1986* (Paris: OECD, 1987), pp. 64–65; R. Schonknecht, J. Lusch, M. Schelzel, and H. Obenaus, *Ships and Shipping of Tomorrow* (Centreville, MD: Cornell Maritime Press, 1983), p. 56; R. F. Gibney, *World Wide Container Data: 1985* (Dalry, Scotland: Container Data, Ltd., October 1985); G. Joseph, "Container Trade Rise Slowdown Forecast," *The Journal of Commerce*, September 14, 1988.

7. "Jumbo Fleet to Grow by 60%," *Container Insight*, Report No. 17 (July 1988):7.

8. The process of jumbo-izing containerships—cutting the vessels apart athwartships and adding new sections to increase their length and carrying capacity—had been developed earlier. For example, in 1985 Hapag-Lloyd jumbo-ized four vessels, increasing their carrying capacity from about 1,760 teus to about 2,600 teus; *The Journal of Commerce*, October 1, 1985.

9. In this regard, the behavior of state-owned shipping companies from countries with non-market economies may be a particular concern. For example, COSCO (Chinese Ocean Shipping Company) has been particularly active in acquiring older containerships. If its containership fleet and those of the Soviet and Polish lines penetrate deeply into the mainstream routes (particularly as cross traders), the imbalance in supply and demand we project could be exacerbated. In addition, if these fleets are used to generate hard currency earnings in accord with national economic objectives and are not required to meet market tests of cost recovery and profitability, the rates they offer might undercut the rest of the industry, exerting downward pressure on rates and earnings.

10. E. E. Toll, "Gradual Tanker Fleet Renewal Urged," *The Journal of Commerce*, June 5, 1989; J. Porter, "How to Replace Aging World Fleet," *The Journal of Commerce*, May 22, 1989; E. Unsworth, Shipbuilding Study Forecast Increases In Demand, Prices," *The Journal of Commerce*, May 3, 1989; "Jumbo fleet may rise 45% by 1990," *Container Insight*, Report No. 16 (April 1988):15.

11. At this writing, however, the Transpacific Stabilization Agreement (as it is currently being called) continues, and a similar initiative involving conference and non-conference carriers in

the Far East-Europe trade is under discussion; J. Porter and J. Jacobs, "Box Lines Agree to Cut Capacity On the Pacific," *The Journal of Commerce,* November 21, 1989. For material on shipper resistance, see J. Porter, "Shippers to Flight Rate Stabilization Efforts," *The Journal of Commerce,* November 22, 1989. Sources on the original agreement include H. Takahashi to M. Chadwin, September 5, 1989 (private correspondence); H. Yamada, "In Celebration of the Birth of TDA," a paper delivered at the International Cargo Handling and Containerization Association meeting, Stockholm, May 31, 1989; M. Magnier, "Ship Lines in Pacific Question Crisis Plan," *The Journal of Commerce,* November 14, 1988; M. Magnier, "Many Trans-Pacific Shippers Say Capacity Cut of 10% Inadequate," *The Journal of Commerce,* October 25, 1988; see also, R. Pearson, *Container Ships and Shipping* (Surrey: Fairplay, 1988).

12. *Container Insight,* Report No. 14 (December 1987): 18–19; it could be argued that the Transpacific Discussion Agreement capacity cutback was, in fact, the first evidence of acceptance of this suggestion by the industry. However, to date it apparently has resulted in a capacity reduction of only about 10 percent among the participants in the main Pacific trades.

13. These relationships may be summarized as follows: materials costs for construction tend to rise in *proportion* to increases in surface area, that is, increases in length, depth, and breadth. However, volume (and hence carrying capacity) tends to be a *cube* function of increases in any of those three dimensions. Furthermore, vessel equipment costs tend to increase more slowly than ship's size, and crew size is largely independent of ship's size. Propulsion power requirements rise at about the *square* of an increase in any of the vessel's three dimensions. However, the power needed to increase the speed of a vessel (whose dimensions are not changed) is roughly a *cube* function of that increase in speed. This is true for speed increases up to vessel's theoretical maximum speed—the point at which the hull's wave resistance makes any additional speed increases prohibitive in terms of the additional propulsive power that is required. For a 1,000-foot-long containership, theoretical maximum speed is in the 35–40 knot range.

14. J. Davies, "West Coast Ports Push Asia-Europe Landbridge," *The Journal of Commerce,* June 19, 1989.

15. Analysis of Corps of Engineers data on the actual depth of jumbo containerships entering their first port of call on a range suggests they almost always ride at least several feet higher than their design draft. Among the reasons are: (1) few vessels sail loaded 100 percent; (2) some of the containers they carry are empties or contain lightweight cargo; and (3) their fuel tanks are almost always partly empty. For related materials, see Richard Schultz, "Deep Draft Vessel Fleet Study, Appendix I and Selected Tables," presented at the Conference on Deep Draft Navigation Data Systems, U.S. Corps of Engineers, Casey Building, Fort Belvoir, Virginia, April 16–17, 1986.

16. This section and the one that follows rely heavily on the remarkable descriptions, analyses, and drawings in *Ships and Shipping of Tomorrow* by Schonknecht and his colleagues, which was cited earlier. Although published in 1983, it remains the best single source on this subject of which we are aware.

17. Some current generation car carriers illustrate the point. They have stern ramps as well as side ramps which permit their cargo of automobiles to be loaded and unloaded faster, even at conventionally configured berths.

18. One of the authors recently observed five ship-to-shore cranes simultaneously deployed on the same jumbo containership at a New Jersey container terminal. If this were done with high-speed cranes, rates higher than 200 moves per hour could be attained for significant portions of the loading and discharging of a single vessel. In order to obtain a comparable rates, a vessel would have to take on and discharge 8 to 10 megacontainers an hour, if each megacontainer held about 25 40-footers.

19. J. Davies, "Designer Lauds Concept of Super-Barge Carrier, *The Journal of Commerce,* October 4, 1988; J. Davies, "Experts Dispute the Validity of Super-Barge Concept," *The Journal*

of Commerce, October 5, 1988; "1-Box Ship Idea Endures," *The Journal of Commerce,* September 12, 1989.

20. E. Unsworth, "Ship Line to Extend Reduced Crew Test," *The Journal of Commerce,* March 8, 1989; D. C. Becker, "Publisher's Notebook," *The Journal of Commerce,* March 22, 1989; B. Vail, "Crew Cuts Please Ship Lines But Take Toll on Seafarers, *The Journal of Commerce,* November 28, 1988; V. W. Stove, "Futuristic Ships Worry Insurers," *The Journal of Commerce,* September 22, 1988.

21. According to C. A. Kane, vice president, industrial engineering for Matson Terminals Inc., two systems were installed, one in Los Angeles and one in Richmond on San Francisco Bay. Each consisted of three pieces of equipment, one of which was a conventional ship-to-shore crane. The second was a "container conveyor," a large platform on wheels that rolled along under the crane as it moved from hatch to hatch and could accommodate up to five containers at once. The third was a huge yard gantry crane that spanned 20 rows of stacks next to the wharf and that shuttled containers between the platform and the stack. The equipment in Richmond is not currently used owing to an absence of containership calls. In Los Angeles the equipment continues to be used, but the automated, computer-driven stacking and inventory control system that was part of the original concept of the yard crane has not been fully developed at this writing.

22. For a description of the new ECT system, see T. M. Hawley, "Rotterdam: Dock Around the Clock," *Oceanus* 32, no. 3 (1989).

23. J. M. Pisani, "Port Development in the United States: Status, Issues and Outlook" (a paper presented at the meeting of the International Association of Ports and Harbors, Miami Beach, Florida, April 22–28, 1989): 37–40.

24. In 1986 an economic impact study of the Port of Hampton Roads estimated that container cargo generated one job for every 239 tons, and breakbulk cargo generated one job for every 175 tons; G. R. Yochum and V. B. Agarwal, *The Economic Impact of Virginia's Ports on the Commonwealth: 1984* (Norfolk, VA: Old Dominion University Maritime Trade and Transport, 1986): 2.

25. As this is being written, Evergreen's and Sealand's 3,500-teu vessels serving the North Atlantic are only calling at three ports on the North American side. In the case of Sealand, these vessels are stopping at only three ports on the European side as well.

26. See, for example, J. Jacobs, "Far East Ports Adopt Bold New Strategies," April 28, 1989; "Intermodalism Lags in Asia," April 5, 1989. R. F. Gibney projected Kaohsiung, Hong Kong, Singapore, Keelung, Busan, and Manila to rank among the top 10 world ports by the year 2001, along with Rotterdam, Kobe, New York-New Jersey, and Yokohama; *World Wide Container Data 1985* (Dalry, Scotland: Container Data Ltd., October 1985): 26.

27. For example, see P. T. Bangsberg, "Need for Terminals Acute: Thai Box Plan Lures Investors," *The Journal of Commerce,* July 6, 1989; M. Magnier, "APL to operate Thai Box Station: Five-Year Deal at Sattahip," *The Journal of Commerce,* May 9, 1989; and J. Porter, "Port Privatization Ranks Low on UK Government's Agenda," *The Journal of Commerce,* March 17, 1989. Although some carriers have become involved in the management of privatized terminals, the number of *dedicated* container terminals is likely to decline. Even large and financially healthy carriers that would prefer to operate their own exclusive facilities are likely to find the high capital and operating costs burdensome and seek other alternatives.

28. R. O. Goss, "A Policy for Seaports," a lecture at the College of Business and Public Administration, Old Dominion University, Norfolk, Virginia, September 7, 1989.

29. For example, see B. Mongelluzzo, "Los Angeles to Stree Access, Environment," *The Journal of Commerce,* November 21, 1989; B. Mongelluzo, "LA Port Weighs Inland Cargo Site," *The Journal of Commerce,* November 24, 1989; and B. Mongelluzo, "S. California Ports Grappling With Growth," *The Journal of Commerce,* December 22, 1989. As noted in Chapter

2, congestion and environmental pressures (as well as marketing strategies) are likely to drive container receiving, storage and staging functions to lower cost inland locations as well as to new sites along the coast.

30. For example, "Containers: Is the Shine Off the Apple For Intermodalism?" *The Journal of Commerce,* April 24, 1989; R. V. Delaney, "Can Intermodalism Compete?" *The Journal of Commerce,* June 21, 1989; C. Dunlap, "Intermodal Marriage Is an Uneasy One" *The Journal of Commerce,* June 16, 1989; "Why Atlanta's Rails Don't Scare Truckers Yet," *Containerisation International* 23, no. 6 (June 1989):35-41. Although some of the criticism about dependability and customer service and the competency of intermodal service providers may be relevant, much of this commentary does not pertain to intermodal moves with a transoceanic leg. Rather it pertains to the slowness of rail-truck service providers to penetrate land-based, primarily domestic or internal transportation markets that remain dominated by pure trucking firms.

31. Ocean-air moves also may provide an alternative for shippers who want something less expensive than direct air freight but faster than either all-water or landbridging service. At the beginning of the 1990s, for example, a small volume of cargo was moving from East Asia to the U.S. West Coast by ship and thence to Western Europe by plane. The cost was reported to be about 60 percent of direct air service, but savings in transit time were estimated to be as much as six days when compared to landbridging; M. Bergen, "Sea-Air Traffic: A Reasonable Option," *The Journal of Commerce,* November 13, 1989.

32. Although the "road-railer" technology may spread quickly beyond North America, some experts are skeptical about the pace at which the double-stack railcar will be adopted. They point out that in Europe, for example, the cost of necessary modifications of tunnels, bridges and power lines might be prohibitive; M. Magnier, "Transport Executives See Competitive World in 90s," *The Journal of Commerce,* November 15, 1989.

33. Soviet officials reported that the Trans-Siberian Container Service's market share of Europe-Far East container traffic was 5 percent in 1988, roughly half of prior levels. They cited "chronic delays and red tape" and "unstable and excessive transit times" among the reasons for the decline. In June 1989 the Soviets announced plans to cut transit times, increase dependability, and improve shippers' access to information about their cargoes; A. Axelbank, "Trans-Siberian Container Traffic Takes Plunge," *The Journal of Commerce,* February 28, 1989; A. E. Cullison, "Soviets to Revamp Landbridge," *The Journal of Commerce,* June 13, 1989.

34. For example, see J. Davies, "Alaskan Ports Seek Polar Route," *The Journal of Commerce,* February 10, 1989.

35. On chassis pools, see R. T. Sorrow, "Chassis Pools Boost Business," *The Journal of Commerce,* October 13, 1989, and A. Gottschalk, "Is It Now Time to Jump Into the Pool?" *The Journal of Commerce,* June 13, 1989.

36. J. Goldstein, "Box Standards Still Far Off: UN Unit Fails to Reach Accord," *The Journal of Commerce,* February 28, 1989; J. Porter, "EC Seeks Genuine Intermodal Network," *The Journal of Commerce,* May 25, 1989; L. Ryan, "Canada Rails Weigh Box Compatibility," *The Journal of Commerce,* June 13, 1989.

37. For a detailed discussion of these developments, see the entire issue of *Container Insight,* Report No. 19 (February 1989).

Index

air freight, 15, 118, 127, 130
all-wheel operation, 20
American President Lines (APL), 8, 96, 103, 104
Antwerp, 11, 70, 101
arrival rate, 35, 38–44
Asia, 4, 7, 9, 11, 15, 25, 112–115, 118, 119, 126, 128, 130, 134, 135
automated guided vehicles (AGV), 123

Baltimore, 11, 101
bar coding, 64, 65, 68, 76, 77
barge, 2, 9, 10, 13, 15, 19, 22, 25, 41, 42, 75, 118, 119, 121, 131, 135
berthing time, 39
breakbulk, 3, 35, 80, 113, 124, 135

C-days, 59–61, 63, 76
capacity, 4, 8, 11, 35–38, 46–50, 53, 54, 56–58, 61, 62, 67, 71, 73, 82, 85, 89, 112–118, 124, 125, 127, 131, 133, 134
capital, 11, 12, 14, 15, 20, 37, 39, 41–43, 50, 56, 62, 67, 73–75, 118, 124, 128, 130, 136
cargo, 1–15, 17, 18, 19, 20, 22–24, 29, 31, 33, 35–39, 41–43, 47–50, 57, 58, 79, 105, 112, 113, 116, 118, 119, 123–131, 133–135
carriers, 2–12, 14, 15, 18, 19, 22–25, 31–33, 35, 39, 41, 42, 49, 50, 52, 56, 58–61, 65–67, 70, 71, 74, 77, 111–125, 127, 129–131, 135, 136
chassis, 2, 8, 10, 15, 19, 20, 22, 24, 25, 29, 31, 32, 36, 51, 54, 57, 58, 60–64, 66, 68, 70, 72–75, 118, 120–122, 126, 128, 129, 136
Chesapeake Bay, 11
China, 7, 133
chunnel, 128
communications, 5, 15, 33, 70, 72, 117, 124
competition, 7, 8, 11, 12, 14, 15, 36, 115, 124–127, 130, 131
computer, 5, 12, 19, 28–33, 36, 40, 45, 49, 63, 64, 68–70, 72–75, 76, 77, 117, 120, 122–124, 130, 131, 135
computerization, 123, 129
computerized, 29, 31, 34, 68, 71, 111
conference, 5, 8, 9, 17, 57, 75, 77, 115, 134

consignee, 13, 22, 23, 33, 42, 63, 127
containerization, 1–7, 10, 17, 111–113, 119, 124, 127, 128, 131, 134
containership, 7, 8, 10, 11, 20, 22, 25, 32, 33, 35, 39, 45, 49–52, 57, 58, 71, 72, 111, 113–121, 133–135, 79–90, 93–106
COSCO (China Ocean Shipping Company, 133
cost accounting, 19, 38
cost equations, 40, 45
crane, 1–3, 10, 12, 20, 23–25, 35, 37, 46, 50–53, 57, 58, 61, 65–67, 74, 75, 116–120, 122, 123–125, 129, 134, 135
crew, 7, 8, 18, 39, 57, 66, 67, 120, 121, 134, 135
CSX, 9, 15, 111, 130
customs broker, 23
customs service, 33
cycle counting, 63, 76
cycle time, 24, 65–67, 75

Dardanelles, 128
decision maker, 28, 29, 35–37, 45, 49, 124
demand, 8, 35, 112, 114, 126, 129, 131, 133, 134
depreciation, 37, 39, 44, 57
deregulation, 5, 17, 102, 103, 125
double-stack, 11, 15, 103, 104, 116, 128
drayage, 22, 41
dredging, 13, 44, 45, 116, 125, 126
dwell time, 58, 59, 61, 62, 76

East Asia, 112, 114, 118, 119, 126, 128
ECT, 18, 25, 35, 70, 116, 122, 123, 127, 135
EDI (electronic data interchange), 12, 17, 129
efficiency, 5–7, 11–14, 19, 20, 32, 33, 48, 57–59, 61, 63, 65–69, 71–74, 76, 115–118, 120, 122–124, 129–131
electronic communications, 5
empties, 59, 60, 63, 134
environment, 3, 4, 9, 11, 49, 50, 65, 124, 130, 131
environmental, 12–14, 114, 120, 126, 127, 131
Europe, 2, 4, 9, 18, 25, 111–114, 118, 119, 128, 129, 134, 136
European Community (EC), 111, 128, 136

137

Evergreen Maritime Corporation, 7, 9, 88, 101, 103, 114, 130, 135
exponential distribution, 40, 44

Far East, 9, 133, 135, 136
fax machines, 127
Federal Express, 130
Federal Maritime Commission, 17, 103
feeder service, 8–10, 13, 99, 117, 118, 121, 123, 125, 94–96
fixed cost, 37, 38, 43, 44, 50, 51, 52, 53
foreign exchange rates, 4
foreman, 29, 32
France, 128
freight consolidator, 22
freight forwarder, 22

gate side, 42, 43
geared, 10, 79
gearless, 10, 79, 114, 117, 118
gentrification, 126, 131
Gibney, 115, 116, 133, 135
government, 5, 8, 12–14, 17, 22, 23, 35, 42, 44, 45, 49, 56, 70, 74, 114, 115, 125, 126, 129, 136

Hamburg, 31, 59, 61, 65, 70, 101
Hampton Roads, 11, 18, 19, 74, 87, 101, 135
Hapag Lloyd, 114
HHLA, 31
hinterland, 11–13, 25, 101
holding area, 41, 71
holding cost, 42, 43
hovercraft, 119
hydrofoils, 119, 131

industrialized countries, 2–4, 7, 121, 125
information technology, 2, 5, 6, 12
inland carrier, 22, 33, 35, 41, 49, 129
inspection, 5, 30, 53, 123, 131
interchange, 12, 17, 22, 28–31, 37, 41, 42, 53, 65, 68, 74
interest rate, 42
intermodalism, 2–6, 8, 14, 15, 96, 103, 111, 112, 127, 129, 131, 135, 136
inventory, 3, 4, 19, 20, 35, 42, 43, 45, 57, 61–64, 76, 77, 97, 129, 135
investment, 3, 4, 10–14, 18, 20, 43, 47, 57, 58, 72, 75, 112, 119, 120, 122, 125, 126, 128, 130
ISO (International Standards Organization), 24, 129

Japan, 3, 4, 7, 8, 115, 121
JIT, 4, 5, 64, 129
jumbo, 10, 18, 114, 116–118, 120, 131, 133, 134

Kobe, 127, 135
Koper, 31
Korea, 7, 8

labor, 1, 3, 6, 7, 18, 20, 35, 37, 41, 50–53, 57, 67, 71–74, 113, 120, 121, 124, 131
landbridging, 8, 9, 12, 15, 18, 116, 125, 128, 134, 136
LASH, 119
less developed countries, 113
line balancing, 57, 65, 66, 76
load center, 10, 12, 14, 118, 123, 125, 131
LO-LO, 82
Long Beach, 116, 126
longshore unions, 6
Los Angeles, 96, 101, 116, 126, 135
LUF-system, 122

management information system (MIS), 20, 28, 29, 68
Maritime Administration, 83
marine terminal, 4–6, 10, 11, 13–15, 19, 20, 22, 32, 34, 35, 36, 41, 46–50, 52, 56–58, 70, 72, 79, 121
marketing, 4, 8, 12–14, 25, 133
Matson Lines, 122
McLean, Malcolm, 1, 10, 14
megacontainers, 119, 120, 122, 131, 135
Merzario/Mantovani, 130
minilandbridge, 96, 125
MIS, 20, 25, 28, 29, 31, 33, 34, 62, 65, 68, 71
model, 18, 36, 37, 39–41, 43–45, 47, 49, 50, 56–58, 73–75, 77, 131
modelling, 38, 40, 42, 43, 75, 76
multi-porting, 9

Nedlloyd, 7, 9, 101, 114, 115, 130
networks, 4, 8, 19, 58, 74, 77, 93–106, 136
New Orleans, 70
New York-NewJersey, Port of, 6
newly industrialized countries, 4, 7, 126
Norfolk, 70, 101
North Africa, 128
North America, 2, 9, 111, 112, 114, 115
North Carolina, 1, 25
nuclear power, 118
NVOCC, 22
NYK (Nippon Yusen Kaisha Lines), 130

officers, 7, 22, 121
Old Dominion University, 17, 18, 74, 135, 136
OOCL (Overseas Orient container Lines), 7

packaging, 3
Panagakos, 74
Panama Canal, 8, 10, 116, 128

pilferage, 3
Pilotage, 35
planes, 127, 129
Poisson distribution, 40, 44
Polish Ocean Lines, 7
port authorities, 6, 13, 14, 71, 125, 126, 131
PRC (China, Peoples' Republic of), 112
pre-stowing, 22
probability distribution, 40
productivity, 6, 41, 117, 124, 126
profit, 9, 118, 132

quay side, 43
queueing, 39, 40, 41, 45, 48, 56
quick response, 4, 129

rail, 1, 2, 9, 11–13, 15, 18, 25, 32, 33, 35, 41, 42, 49, 52, 71, 72, 127, 128, 136
railcars, 2, 13, 116
railroad, 1, 4, 5, 14, 15, 22, 32, 33, 41, 50, 64, 71, 72, 128
reach stackers, 24
recipient, 1, 22, 33
refrigeration, 133
registries, 8, 121
RO-RO, 7, 10, 80–82, 89, 117–119, 121, 131
Rotterdam, 11, 25, 70, 101, 116, 135
round-the-world service, 9, 116
routes, 4, 8, 9, 10, 75, 113, 114, 117–119, 121, 123, 125, 128, 131, 133, 136

Seabee, 119
Sealand, 1, 2, 7, 9, 15, 18, 103, 115, 123, 129, 130, 135
semi-container, 10, 80, 89
Senator Lines, 9, 101
service time, 39–41, 43, 57
ship building, 112, 114
ship-to-shore crane, 65, 66, 89
ship's agent, 23
shipper, 3, 5, 6, 11–15, 18, 20, 22–25, 29, 33, 35, 41–43, 45, 49, 56, 59, 60, 63, 64, 68, 73, 74, 120, 122, 123, 127, 129, 134, 136
Shipping Act of 1984, 5, 8, 17, 103
simulation, 19, 36, 40, 45, 48, 49, 57, 58, 73–77
Singapore, 7, 101, 116, 135
Soviet Union, 112
space utilization, 57–60, 63, 73, 76
spreader bar, 66
stack, 8, 10, 11, 15, 19, 20, 24, 25, 31–33, 36, 58–61, 63–68, 70–74, 116, 121–123, 127, 128, 135
statistics, 45
stevedore, 20, 22, 23, 33, 72, 73, 87
stowage, 5, 19, 23, 25, 32, 38, 67, 123, 124

stowage planning, 5, 19, 23, 123, 124
straddle carrier, 24, 25, 31, 32, 58, 60, 61, 65–67, 70, 77, 122, 123
strategic management, 130
stuffing and stripping, 6, 125
submarines, 119
subsidies, 17, 44
SWATH, 120
Sweden, 7, 18, 120

Taiwan, 7, 8
tanker, 2, 114, 119, 133
terminal, 2, 4–6, 9–15, 19–25, 28–39, 41–54, 56–59, 111, 112, 116–126, 128, 130, 131, 133, 135, 136
Terminal Consulting Holland, 75
terminal manager, 57, 63, 65
terminal operator, 6, 13, 22, 23, 33, 35–38, 46, 49, 50, 56–59, 67, 71, 122, 130
terminal worker, 19, 20, 124
time-and motion study, 65, 76
throughput, 32, 36–38, 46–49, 57, 58, 62, 64, 69
Tokyo, 17, 127
Transatlantic Shipping Company, 120, 130
transient, 58–60
Transpacific Discussion Agreement, 115, 134
transtainer, 20, 24, 25, 31, 32, 51–53, 58, 60, 61, 65–67, 70, 71, 122
truck, 1, 2, 4, 9, 13, 15, 19, 20, 22, 24, 25, 35, 41, 42, 49, 50, 52, 66, 121, 128, 136
trucker, 19
trucking, 1, 5, 136
tugboat, 35, 74

unit train, 9, 12, 13, 15, 25, 32, 33, 72, 116, 121, 127, 128
United Kingdom, 7
United States, 3–13, 15, 17, 18, 23, 25, 31, 48, 56, 57, 68, 77, 102, 103, 112, 115, 116, 118, 122, 127, 128, 129, 135
UPS (United Parcel Service), 130
U.S. Lines, 9, 88, 96

variable cost, 37, 38, 43, 44, 50, 51, 53, 72
Virginia, 17–19, 25, 74, 77, 134–136
Virginia Center for World Trade, 17, 19, 74

waiting time, 39
Waltershof, 59, 61, 65
wharfside crane, 50, 66, 67, 80
work preservation, 6

yard equipment, 12, 20, 24, 116, 123
yard hustler, 20, 58
Yugoslavia, 31

About the Authors

Mark Lincoln Chadwin is a professor of Management at Old Dominion University. Since coming to the university in 1980, he has been director of Business Research, director of the Maritime Trade and Transport program, and director of research in the Virginia Center for World Trade. Previously he was executive director of the Illinois Economic and Fiscal Commission, a senior research associate at The Urban Institute in Washington, DC, national security advisor to Senator Birch E. Bayh, Jr., and an assistant to W. Averell Harriman. Dr. Chadwin is the author of books, reports, and articles that range across diplomatic history, contemporary foreign policy, evaluation research, organization design, program implementation, and international shipping and trade. He holds degrees from Yale and Columbia.

James A. Pope is an associate professor at Old Dominion University and chairman of the Department of Management Information Systems/Decision Sciences. He has been director of the university's Bureau of Business Research and a research fellow in the Maritime Trade and Transport program. The author of numerous articles and research reports on logistics, marine terminal operations, and simulation techniques, Dr. Pope was an officer in the United States Air Force and taught at Guilford College, the University of North Carolina, and the University of Colorado (Denver), before coming to Old Dominion in 1980. He has degrees from the College of Wooster, Northwestern University, and the University of North Carolina, and is a Certified Fellow in Production and Inventory Management (CFPIM).

Wayne K. Talley is an Eminent Scholar at Old Dominion University and chairman of the Department of Economics. A member of the university's faculty since 1972, he has also been a faculty fellow at the U.S. Department of Transportation's Transportation Systems Center, an industry economist at the Interstate Commerce Commission, a visiting fellow of the Transport Studies Unit at Oxford University, a visiting research fellow of the Centre for Transport Policy Analysis at the University of Wollongong, and a research fellow of Old Dominion's Maritime Trade and Transport program. Dr. Talley has written extensively on transportation cost analysis and is the author of two recent books, *Transport Carrier Costing* and *Introduction to Transportation*. He holds degrees from the University of Richmond and the University of Kentucky.